Beryllium
Environmental Analysis and Monitoring

Beryllium
Environmental Analysis and Monitoring

Edited by

Michael J. Brisson and Amy A. Ekechukwu
Savannah River Nuclear Solutions, Savannah River Site, Aiken, SC, USA

RSC Publishing

ISBN: 978-1-84755-903-6

A catalogue record for this book is available from the British Library

© Royal Society of Chemistry 2009

Published by The Royal Society of Chemistry,
Thomas Graham House, Science Park, Milton Road,
Cambridge CB4 0WF, UK

Registered Charity Number 207890

For further information see our web site at www.rsc.org

Preface

Beryllium is a metal with unique properties that make it useful for a number of applications, from consumer products such as cell phones, to nuclear weapons components. These unique properties make it difficult to find alternatives to beryllium and ensure that it will continue to be used for the foreseeable future. However, for some individuals, exposure to beryllium particulates in the workplace can lead to a sensitization reaction. Sensitized individuals with beryllium particulates in the lungs are at risk for chronic beryllium disease (CBD), which can have a long latency period before symptoms appear. Sensitization and/or disease can result from exposure at very low levels. As a result, control of exposures to beryllium in the workplace is essential. Although engineering controls are normally the first line of defense, exposure monitoring, including sampling and analysis, is also important and is typically mandated by regulation.

While most metals and metalloids have occupational exposure limits in the range of milligrams per cubic metre, limits for beryllium are in the microgram or sub-microgram per cubic metre range. Additionally, some forms of beryllium in the workplace are highly refractory, making them difficult to dissolve for analytical purposes. These considerations pose unique challenges for monitoring of beryllium exposure in the workplace. Some of the challenges include: sampling a sufficient air volume to evaluate short-term exposures; sampling settled dust (in some cases accumulated over decades) on a wide variety of surfaces; preparing samples to ensure that all of the workplace beryllium forms are detected; and obtaining sufficient analytical sensitivity. Since datasets often have a large percentage of results below the laboratory's reporting limit, data reporting itself is often a challenge.

Although there is now considerable information on beryllium sampling and analysis in the literature, much of it within the last decade, there has up to now been no single compendium to survey the literature and provide guidance on best practice. Providing such a resource is our goal for this book. We do not

Beryllium: Environmental Analysis and Monitoring
Edited by Michael J. Brisson and Amy A. Ekechukwu
© Royal Society of Chemistry 2009
Published by the Royal Society of Chemistry, www.rsc.org

promote a one-size-fits-all approach; instead, our goal is to provide information that will enable users to ensure that their sampling and analysis techniques are fit-for-purpose. Hopefully, we will promote more consistency along the way.

There are likely more challenges to come. Since there is no known exposure–response relationship for beryllium sensitization or disease, the trend toward lower occupational exposure limits may continue indefinitely. There remains some difference of opinion on the need for particle size-selective sampling, and what fractions should be sampled. We also do not know whether some anthropogenic forms of beryllium are more toxic than others. Future information may point to a need to differentiate, say, beryllium oxide from beryllium metal or alloy. While major research laboratories can do that today, the typical industrial hygiene laboratory cannot. New information on these topics will hopefully spawn improvements in the areas covered in this book. In the meantime, we present the state of the art as it is today and trust it will be of benefit throughout the scientific community.

Michael J. Brisson
Amy A. Ekechukwu
Co-editors

Contents

Beryllium: Environmental Analysis and Monitoring
Edited by Michael J. Brisson and Amy A. Ekechukwu
© Royal Society of Chemistry 2009
Published by the Royal Society of Chemistry, www.rsc.org

CHAPTER 1

Overview of Beryllium Sampling and Analysis*‡

Occupational Hygiene and Environmental Applications

MICHAEL J. BRISSON

Senior Technical Advisor, Savannah River Nuclear Solutions, Analytical Laboratories, Savannah River Site, Aiken, SC 29808, USA

Abstract

Because of its unique properties as a lightweight metal with high tensile strength, beryllium is widely used in applications including cell phones, golf clubs, aerospace, and nuclear weapons. Beryllium is also encountered in industries such as aluminium manufacturing, and in environmental remediation projects. Workplace exposure to beryllium particulates is a growing concern, as exposure to minute quantities of anthropogenic forms of beryllium may lead to sensitization and to chronic beryllium disease, which can be fatal and for which no cure is currently known. Furthermore, there is no known

---

*This article was prepared by a US Government contractor employee as part of his official duties. The US Government retains a nonexclusive, paid-up, irrevocable license to publish or reproduce this work, or allow others to do so for US Government purposes.

‡ *Disclaimer*: Mention of company names or products does not constitute endorsement by Savannah River Nuclear Solutions (SRNS) or the US Department of Energy (DOE). The findings and conclusions presented in this chapter are those of the author and do not necessarily represent the views of SRNS or DOE.

Beryllium: Environmental Analysis and Monitoring
Edited by Michael J. Brisson and Amy A. Ekechukwu
© Royal Society of Chemistry 2009
Published by the Royal Society of Chemistry, www.rsc.org

exposure-response relationship with which to establish a "safe" maximum level of beryllium exposure. As a result, the current trend is toward ever lower occupational exposure limits, which in turn make exposure assessment, both in terms of sampling and analysis, more challenging. The problems are exacerbated by difficulties in sample preparation for refractory forms of beryllium, such as beryllium oxide, and by indications that some beryllium forms may be more toxic than others. This chapter provides an overview of sources and uses of beryllium, health risks, and occupational exposure limits. It also provides a general overview of sampling, analysis, and data evaluation issues that will be explored in greater depth in the remaining chapters. The goal of this book is to provide a comprehensive resource to aid personnel in a wide variety of disciplines in selecting sampling and analysis methods that will facilitate informed decision-making in workplace and environmental settings.

1.1   Introduction

Control of occupational exposure in the workplace, characterization of environments or legacy areas, and management of environmental or workplace remediation projects, all require careful planning and execution, including development of appropriate sampling plans, up-front understanding of laboratory capabilities, and proper evaluation of analytical data. This involves a number of disciplines, including industrial hygienists, laboratory personnel, statisticians, and line management. Even before a sampling plan is developed, additional disciplines such as medicine, immunology, toxicology, and epidemiology, are involved to tell us the health risks of the material we are trying to control. Additional disciplines, such as engineering, assist us with implementing the full hierarchy of controls,[1] of which sampling and analysis are a part, to minimize exposure to toxic substances in workplace and environmental settings. All of these disciplines must work closely together, beginning with the design phases of a project or facility, through the end of a project's lifecycle, to ensure an outcome that protects workers but also avoids unnecessary costs to the project.

Perhaps nowhere is this more true than with beryllium. Because beryllium exposure must be managed at ultra-trace levels (with the trend being toward even lower levels), the sampling and analytical challenges associated with measuring beryllium are greater than for most other metal or metalloid particulates. This includes workplaces actively using beryllium, legacy areas where beryllium was used in the past, and environmental remediation projects. New facilities where beryllium will be used need to be designed not only with appropriate engineering controls, but also with consideration of beryllium sampling and analytical requirements.

This book provides information on sampling and analysis techniques that have been developed to ensure that beryllium particulate (whether in natural or anthropogenic forms) can be effectively sampled and analyzed, and the resulting data properly evaluated for sound decision-making in workplace and environmental settings. This book is not intended to provide detailed medical or toxicological information, nor does it discuss engineering controls. It is focused primarily on the sampling and analytical state-of-the-art.

This chapter provides background information on beryllium sources, uses, health risks, and exposure limits. It then provides an overview of sampling and analysis issues to set the stage for the detailed discussion of these issues and techniques in the chapters to follow.

1.2 Goals of this Book

The primary goal of this book is to be a resource that can be used by all of the disciplines involved in beryllium health and safety management, to enable the best possible sampling and analytical decision-making so that workers are better protected from the risks of beryllium in the workplace. Its primary users would include industrial hygiene practitioners, analytical laboratory personnel, statisticians, and managers of projects or processes that either utilize beryllium or characterize beryllium in legacy or environmental settings. This book should help such users understand current capabilities and limitations in beryllium sampling and analysis, both in their own disciplines and in the others, and the need for good communication with other disciplines to assure success. It is also hoped that this book will be useful in academic, research and development settings to encourage additional research to address the many limitations in our current understanding and capabilities.

It is not the intention of this book to tell users to sample or analyze by some prescribed method(s). There is no "one size fits all" approach to beryllium sampling and analysis, but it is important that selected methods be fit for purpose and be defensible (as applicable) to customers, regulators, accrediting agencies, managers, and perhaps most importantly, to workers whose beryllium exposures are being characterized and managed.

1.3 Background

Beryllium (atomic number 4) is a lightweight metal (density 1.85) with a high melting point (1287 °C), stiffness (Young's modulus 287 GPa) and thermal conductivity (190 W m^{-1} K^{-1}).[2,3] These properties make beryllium a highly desirable component for a wide variety of applications.

1.3.1 Beryllium Sources

Beryllium occurs naturally in some 30 different mineral species.[3] In the Earth's crust, beryllium content is estimated at 2–5 parts per million (ppm) overall, with specific rocks having up to 15 ppm.[4] For the extraction of elemental beryllium, the species of importance are the beryllium alumino-silicate mineral beryl ($Be_3Al_2Si_6O_{18}$) and the beryllium silicate hydroxide mineral bertrandite [$Be_4Si_2O_7(OH)_2$], with bertrandite as the principal mineral mined in the United States, and beryl the principal mineral in other countries.[4] Beryl is roasted with

Table 1.1 Beryllium content in various substances.[4]

Substance	Beryllium Content
Coal	$1.8–2.2\,E^{+06}\,\mu g\,kg^{-1}$
Coal ash	$4.6\,E^{+07}\,\mu g\,kg^{-1}$
Stack emissions from coal-fired power plants	$0.8\,\mu g\,m^{-3}$
Cigarettes	ND–$0.74\,\mu g$ per cigarette
Fertilizers	$<200–13\,500\,\mu g\,kg^{-1}$
US drinking water	$0.5\,\mu g\,L^{-1}$
Air (US average)	$<3\,E^{-05}\,\mu g\,m^{-3}$
Kidney beans	$2200\,\mu g\,kg^{-1}$
Crisp bread	$112\,\mu g\,kg^{-1}$
Garden peas	$109\,\mu g\,kg^{-1}$

sodium hexafluorosilicate to form beryllium fluoride, which is water-soluble. From the fluoride, beryllium may be precipitated as beryllium hydroxide by adjusting the pH to 12, or may be obtained by reduction of the fluoride with magnesium.[2] For bertrandite, the ore is leached with sulfuric acid; solvent extraction of the sulfate solutions ultimately produces beryllium hydroxide.[5] In 2007, active mine production was principally in the United States, China, and Mozambique, with minor amounts elsewhere.[6]

Beryllium also is found in bauxite ore used in the manufacture of aluminium. The amount of beryllium varies with the source of the bauxite. While bauxite is not a beryllium source for production purposes, aluminium smelter workers can be exposed to beryllium in pot emissions.[7,8] Table 1.1 contains additional data on beryllium in a variety of materials based on information from the US Agency for Toxic Substances and Disease Registry (ATSDR).[9]

1.3.2 Beryllium Uses

1.3.2.1 Beryllium metal

Beryllium metal is used in nuclear weapons, aircraft brake parts, spacecraft structures, navigation systems, X-ray windows, mirrors, and audio components.[9] The metal is also a neutron reflector used in nuclear reactors.

1.3.2.2 Beryllium alloys

Beryllium alloys represent the largest use of beryllium. Copper–beryllium alloys typically have 0.15–2.0% beryllium content,[5] and are widely used because they exhibit good conductivity, are resistant to corrosion, have high hardness, and are non-magnetic. Copper–beryllium alloys are used for applications such as coaxial connectors in cell phones, computers, aircraft bushings, non-sparking tools, automotive switches and sensors, and plastic injection molds.[5,9] Aluminium–beryllium alloys, such as Brush-Wellman's AlBeMet®, are used as optical substrates for night vision systems and avionics applications.[10] Nickel–beryllium alloys have good spring characteristics and are used in applications such as thermostats and bellows.[11]

1.3.2.3 Beryllium oxide

Beryllium oxide is used in a variety of ceramics applications such as medical laser bores, integrated circuits, electronic heat sinks and insulators, microwave oven components, gyroscopes, and thermocouple tubing.[9]

1.3.3 Health Risks

The most noticeable adverse health effects from beryllium exposure are those affecting the respiratory system; however, effects on the lymph nodes, skin, and other target organs have been documented.[12]

Acute beryllium disease is an inflammation of the entire respiratory tract caused by exposure to high levels of soluble beryllium.[9] Symptoms may range from mild nasopharyngitis to severe pneumonitis, which could be fatal. These effects were reported in the US in the 1940s.[13,14] All cases in the 1948 study[13] involved exposures greater than $0.1\,mg\,m^{-3}$. Imposition of exposure limits after 1950 all but eliminated acute beryllium disease.

At significantly lower levels, exposure to airborne beryllium particulate can cause an immune system response known as *beryllium sensitization* (BeS).[12] Estimates of BeS range from 0.9% to 21.4% of those exposed,[15] with some industrial processes having a higher prevalence of BeS than others. There is no established dose–response relationship, but BeS has been attributed in some studies to exposures below $0.2\,\mu g\,m^{-3}$ (mean daily lifetime weighted average).[16–18] Studies are ongoing as to the mechanism by which sensitization occurs, but genetic susceptibility is believed to be a factor.[18,19]

Sensitized individuals may go on to develop *chronic beryllium disease* (CBD), a debilitating and potentially fatal lung disease characterized by lesions in the lung known as granulomas.[20] Because the mechanism of progression from exposure to BeS to CBD is not well understood, it is possible that once an individual is sensitized, a risk of developing CBD exists even if there is no further exposure to beryllium.[20] Also, recent studies suggest that dermal exposure, in addition to causing contact dermatitis in some workers,[15,21] may also be a pathway to BeS,[22,23] although CBD appears to require some pulmonary exposure.[12] Controlling workplace exposures to prevent BeS and/or CBD is the primary driver for the sampling and analysis activities described in this book.

Thus far, cases of CBD have involved exposure to anthropogenic forms of beryllium, *i.e.* metal, alloy, or oxide. Exposure to natural forms of beryllium (beryl or bertrandite) has not been shown to result in CBD, although BeS has been reported from such exposure.[12]

Additionally, the International Agency for Research on Cancer (IARC) has determined that there is sufficient evidence that beryllium and compounds are human carcinogens.[24] The US National Toxicology Program has reached a similar conclusion.[25] Alternative conclusions have been presented in the literature, and discussion of the differing positions was ongoing at the time of writing.[26–29]

1.3.4 Occupational Exposure Limits

In the US, initial exposure limits were established based on studies in the late 1940s by the Atomic Energy Commission.[30] By this point, the existence of CBD and the need to protect against it, as well as acute beryllium disease, had been established. Additionally, instances of CBD were reported among residents near the beryllium plant in Lorain, Ohio. The initial proposal was for a peak exposure limit of $25\,\mu g\,m^{-3}$, intended as protection against acute disease. Next, an ambient air limit of $0.01\,\mu g\,m^{-3}$ was adopted for community protection. Finally, a limit value of $2\,\mu g\,m^{-3}$ was proposed as an eight-hour time-weighted average to protect against CBD. This proposal was based on an extrapolation of the prevailing limit value for heavy metals such as arsenic and lead, accounting for the lower atomic weight of beryllium.

Within the US, two of the three original limits remain in place at the time of writing. The limit value of $2\,\mu g\,m^{-3}$ (eight-hour time-weighted average) remains in place as the permissible exposure level (PEL) of the US Occupational Health and Safety Administration (OSHA).[31] This limit value is also in use in many other countries, and until very recently, was also the threshold limit value (TLV®) of the American Conference of Governmental Industrial Hygienists (ACGIH®). The original ambient air quality standard of $0.01\,mg\,m^{-3}$ also remains.[32] In 1997, ACGIH® adopted a short-term exposure limit (STEL) of $10\,mg\,m^{-3}$.[33] Additional discussion on limit values in air can be found in Chapter 2.

A number of studies have established that an occupational exposure limit of $2\,\mu g\,m^{-3}$, as well as the current STEL, are not adequately protective.[16–18,34–35] As a result, proposals have been made to lower these limits. After issuing several notices of intended change (1999, 2005, 2006 and 2007), ACGIH® in early 2009 adopted a new TLV® of $0.05\,mg\,m^{-3}$,[33] but did not adopt a proposal to lower the STEL to $0.2\,mg^{-3}$.[36] OSHA has also begun the process to lower its PEL, possibly to as low as $0.1\,mg\,m^{-3}$.[37] A listing of occupational exposure limits for selected countries is provided in Table 1.2.

1.3.5 Impact of US Department of Energy Regulation

In 1999, the US Department of Energy (DOE) promulgated a regulation known formally as the Chronic Beryllium Disease Prevention Program (informally the Beryllium Rule),[38] which established three action levels for DOE facilities:

(a) An airborne beryllium limit of $0.2\,\mu g\,m^{-3}$
(b) A housekeeping limit of $3.0\,\mu g$ per $100\,cm^2$ for surfaces within beryllium work areas

Table 1.2 International occupational exposure limits for beryllium and compounds.[40]

Country	Limit Value (µg/m) Eight-Hour Time-Weighted Average	Short Term
Austria	2	8
Belgium	2	
Canada (Quebec)	0.15	
Denmark	1	2
France	2	
Hungary		2
Japan	2	
Poland	1 (inorganic compounds)	
Spain	2[a]	
Sweden	2	
Switzerland	2 (inhalable aerosol)	
United Kingdom	2	
United States (OSHA)[b]	2	5

[a]Spain has a separate limit for beryllium oxide (same Limit Value as above).
[b]Changes to the US OSHA limits have been proposed and were pending at the time of writing.

(c) A limit of 0.2 µg per 100 cm² for release of equipment to the public or to "non-beryllium" work areas

All of these action levels are empirical, as DOE recognized that the existing PEL was not adequately protective and, while wanting to take some steps to improve worker protection, did not have an exposure–response relationship on which to base any action levels.

At the time of writing, there is still no exposure–response relationship. DOE did not wait for such a relationship, and its action appears to be part of a trend toward lower empirical exposure limits and action levels. As noted previously, ACGIH® also acted in 1999, issuing the first of several notices of intended change. The 1999 proposal was in fact for a TLV® at the DOE action level of $0.2\,\mu g\,m^{-3}$,[36] with subsequent proposals even lower. In North America, the state of California[39] and the province of Quebec[40] have also lowered their workplace air exposure limits to $0.2\,\mu g\,m^{-3}$ and $0.15\,\mu g\,m^{-3}$, respectively. Finally, in OSHA's report on its preliminary draft standard, options it has considered include essentially adopting the DOE Beryllium Rule.[37]

The DOE Beryllium Rule is presently the only regulation with specific action levels for contaminated surfaces. However, others may soon follow. Studies of surface sampling have been performed in Quebec and at some US Department of Defense sites. OSHA has indicated that a surface PEL is a possible option for its new standard.[37] Finally, a recent US National Academy of Sciences report commissioned by the US Air Force suggests that surface and skin contamination correlate with airborne contamination, and suggests that the Air Force perform surface sampling consistent with the DOE standard.[41]

Thus, it is clear that the DOE Beryllium Rule has had appreciable impact within the US, and it appears reasonable, based on studies such as Day *et al.*[23]

and the National Academy of Sciences (NAS) report,[41] to predict that both surface and dermal sampling for beryllium will increase, at least within the US. Even in the absence of specific numerical surface standards, some degree of surface sampling may be appropriate as part of an overall beryllium house-keeping program. Thus, the discussion in Chapter 3 of techniques for sampling and analysis of surfaces should be beneficial.

1.3.6   Environmental Beryllium and Soil Remediation

Another provision of the DOE Beryllium Rule is that, for purposes of com-plying with action levels, background beryllium levels from soil, if known, may be subtracted.[38] This has led to a need for a reliable method to measure ber-yllium levels in soil. In addition, environmental remediation at sites where beryllium was used in the past have included cleanup goals for beryllium, requiring the ability to measure beryllium levels in soil.[42] A reliable method has been recently developed[43] and is described in Chapter 7.

1.3.7   Beryllium in Water

Sampling and analysis of beryllium in water are outside of the scope of this book. The reader is referred to published sampling and analysis methods such as those published by ASTM International,[44] the US Environmental Protection Agency (USEPA),[45] or *Standard Methods*[46] for more information. For data on levels of beryllium found in ambient water, refer to the ATSDR toxicological profile[9] for more information.

1.4   Sampling Overview

1.4.1   Air Sampling

Workplace atmosphere sampling for beryllium has been taking place for over 60 years. The background, technical basis, and current issues associated with beryllium air sampling are described in detail in Chapter 2, although some of the key issues are given here.

First, it is often challenging to obtain a sample of sufficient air volume. It is typically necessary to sample the breathing zone of the worker, using air pumps that sample at a relatively low rate. For short-duration jobs (less than an hour), the available volume is often less than $0.1 \, \text{m}^3$. When air volume is low, ana-lytical sensitivity must be greater to obtain a meaningful result in relation to an action level or exposure limit.

A second issue is size-selective sampling. In the US, closed face cassette (CFC) sampling is commonly used.[47] This method is ostensibly for "total" dust, although as noted in Chapter 2, larger aerosol particulate is not sampled particularly well with the CFC. Outside the US, the inhalable sampling fraction of ISO 7708[48] is more widely used. ACGIH®, in its 2009 TLV®,[33] adopted the inhalable convention for beryllium as part of a general move in that direction for its TLVs®. This will require a re-evaluation of current sampling techniques for those who choose to utilize the ACGIH® TLV® and are not using an inhalable sampling method.

Finally, it is necessary to note that some beryllium particulate may deposit on interior walls of samplers. There is currently no consensus on whether such wall deposits need to be included in the sample that is analyzed by the laboratory. Subsequent chapters elaborate on this issue as well as on techniques to include wall deposits for those who choose to do so.

As noted previously, the focus of this book is on beryllium sampling and analysis in workplaces, not on environmental beryllium sampling. However, the first ambient air standard adopted in the US was in fact for beryllium.[30] There is, therefore, regular air monitoring across the US for beryllium. ATSDR reports the ambient levels in the US to be $0.03-0.2$ ng m^{-3}, with higher levels in urban areas due to the burning of coal and fuel oil.[9] Comparable results have been obtained in studies in Germany and Japan.[49]

1.4.2 Surface Sampling

Background, technical basis, and current issues associated with beryllium air sampling are described in detail in Chapter 3. Some of the key issues are given here.

Proper planning is essential for successful surface sampling. It is necessary to understand both the nature of the surfaces being sampled (for characteristics such as roughness and porosity) and the characteristics of the dust on the surface (such as oiliness and thickness of dust) to select the correct sampling technique (such as vacuum sampling or surface wiping). It is also necessary to understand the end purposes for the sampling campaign, including required data quality objectives. This information aids in selection of the appropriate number of samples and sampling points. The capabilities of the laboratory performing the analyses must be understood so that the results can be used as intended. Finally, how the data will be evaluated and communicated to the end user (discussed in Chapter 8) should be understood prior to commencing of sampling. In addition to Chapter 3, resources such as the American Industrial Hygiene Association's *A Strategy for Assessing and Managing Occupational Exposures*[50] are useful for planning. While these criteria are also useful for planning air sampling campaigns, they are of particular importance for surface sampling.

Collection efficiency of surface wipes is another important factor. For beryllium surface wiping, there is limited information available in the published literature; one of the few published studies was by Dufay and Archuleta.[51] The prevailing view is that wetted wipes have better collection efficiency than dry wipes, though some unpublished studies have questioned that view. A DOE study showed that selection of wetting agent is important (*e.g.* alcohol-wetted wipes may be better for oily surfaces than water-wetted wipes).[52] Also, some surfaces may be damaged by wet wiping; for these, dry wiping may be the best method available.

Finally the ability of the laboratory to handle the surface wiping matrix should be verified before collecting samples using that matrix. Additional details on effective sample preparation techniques can be found in Chapter 4.

1.4.3 Dermal and Soil Sampling

As noted previously, dermal exposure to beryllium has been recently identified as a potential route to BeS,[22,23] and some studies have suggested a correlation between dermal exposure and airborne beryllium levels.[41] The European Committee for Standardization (CEN) has issued a report and a technical specification on dermal sampling.[53,54] The CEN technical standard describes generic techniques for dermal sampling; however, specific methods for beryllium are not currently available. Task groups have been formed within ASTM International and the International Standards Organization (ISO) to develop dermal sampling methods for beryllium and other contaminants.

Detailed information on soil sampling is outside of the scope of this book. The user is referred to standard methods such as those issued by ASTM International[55] for soil sampling guidance.

1.5 Analysis Overview

1.5.1 Summary of Current Techniques

Analytical techniques commonly used in the US and Europe include inductively coupled plasma atomic emission spectroscopy (ICP-AES), inductively coupled plasma mass spectroscopy (ICP-MS), and atomic absorption spectroscopy (AAS).[56] Additionally, a molecular fluorescence method for beryllium has been recently developed in the US and has demonstrated sensitivity comparable to ICP-MS. These techniques are described in Chapters 6 and 7.

A variety of alternative techniques have been attempted. These include laser-induced breakdown spectroscopy (LIBS),[57] microwave-induced plasma spectroscopy (MIPS),[58,59] aerosol time-of-flight mass spectroscopy (TOFMS),[60] and surface-enhanced Raman spectroscopy.[61] In general, these techniques require significantly less sample preparation than those described in Chapters 6 and 7. However, due to issues with lack of precision at lower analyte levels and with ability to process surface wipes, these methods have not gained wide acceptance.

1.5.2 Sample Preparation

A key consideration in effective analysis of beryllium at trace levels is sample preparation for subsequent analysis by ICP-AES, ICP-MS, AAS, or fluorescence. Sample preparation techniques are described in Chapters 4 and 5. Since analytical techniques used for trace-level beryllium analysis require the beryllium to be in solution, it is of paramount importance that the selected sample preparation method digests or extracts all of the beryllium into the solution used for analysis. As noted in Chapter 4, a number of standard sample preparation techniques are available. A survey of 16 laboratories (primary US DOE) conducted in 2004 indicated not only that a wide variety of methods were being used, but also that most labs found it necessary to modify a "standard" method in some fashion.[62]

Of the forms of beryllium typically encountered in workplace air and surface samples, beryllium oxide (BeO) is the most difficult to bring into solution. Issues affecting the ability of a sample preparation method to dissolve or extract BeO include particle size distribution and BeO firing temperature.[63,64] Until recently, the lack of a BeO reference material hindered the ability to provide a definitive evaluation of digestion and extraction protocols for effectiveness with BeO.[65] In spring 2008, however, the US National Institute of Standards and Technology (NIST) released a BeO Standard Reference Material (SRM).[66] It is hoped that this material will make possible a more effective validation of beryllium sample preparation methods. It is also hoped that the BeO reference material can be used in proficiency testing programs, such as that conducted by the American Industrial Hygiene Association (AIHA), to provide a more robust test of digestion and extraction methods. The current AIHA program is based on beryllium acetate, which is water-soluble and thus easy to bring into solution. A BeO-based proficiency testing program would provide greater assurance that participating laboratories are effective in bringing the various forms of beryllium in workplace samples into solution.

Sample preparation for beryllium in soil is also described in Chapter 4. Beryllium silicates and aluminosilicates typically require a more robust preparation method than does BeO; however, these forms are not typically encountered by industrial hygiene laboratories.

1.5.3 Data Evaluation and Reporting

Another issue that has gained increased attention in recent years is that of statistical evaluation of beryllium analytical data. This is the focus of Chapter 8. In many instances, a majority of data in many datasets consist of non-detects, or values that fall below the laboratory's reporting limit (RL). Such results are typically reported as "<RL", often referred to as "censored data". This form of data censoring is required by accreditation bodies such as AIHA. In these instances, proper evaluation of a dataset can be very difficult. If accredited data are not required, reporting of data below the laboratory limit, with appropriate caveats, is one option discussed in Chapter 8. This chapter also discusses issues with reporting data to downstream customers, stakeholders, and affected workers.

1.5.4 Future Analytical Challenges

As mentioned previously, the trend toward lower occupational exposure limits for beryllium will create greater challenges for sample preparation and analysis.

Already, the current DOE action levels for airborne and surface contamination are pushing the limits of techniques such as ICP-AES, which is the most commonly used in US industrial hygiene laboratories. Proposals such as the Short Term Exposure Limit (STEL) of $0.2\,\mu g\,m^{-3}$ that had been proposed by ACGIH®, may bring about requirements for sensitivity beyond the reach of ICP-AES. As an example, a 15-minute air sample using a $2\,L\,min^{-1}$ personal air pump would provide $30\,L$, or $0.03\,m^3$, of air. Multiplying the STEL by this volume results in a *de facto* sensitivity requirement of 6 ng per sample. However, it is considered good laboratory practice to have a reporting limit of one-tenth the action level, which in this case translates to 0.6 ng per sample. For soluble forms of beryllium, requiring only small amounts of solution, analysis at these levels by ICP-AES has been reported,[67] but for non-soluble forms of beryllium, analysis at these levels is likely beyond the reach of ICP-AES due to the higher dilution factors required.

An additional challenge is the need for faster analysis, preferably in or near real time, while retaining excellent sensitivity. Most laboratories can analyze "rush samples" within a few hours; however, for routine samples, a time of 24 hours or longer is more typical. This lag time is driven by competition from other samples and, in many cases, by the sheer volume of beryllium samples, which for some labs is in the tens of thousands per year.[62] Radiologically contaminated samples, which represented 19% of the total in the 2004 survey, take longer and are more expensive. The expense (millions of US dollars for DOE alone) is another reason why real-time, or near real-time, beryllium monitoring would be desirable.

In the first few years following implementation of the DOE Beryllium Rule,[38] several attempts were made, focusing on direct-solids measurement techniques that did not require solubilization of the beryllium and thus could be expected to provide faster results and could be deployed closer to field locations. However, when these initial efforts were not successful, it became evident that more costly and time-consuming research and development would be necessary to develop suitable (near) real-time monitoring equipment. To date, this level of resource commitment has not been available. Optimization of existing sampling and analysis methods seems the principal path in the near term for improving our existing capabilities.

Acknowledgements

We appreciate the contributions of the Beryllium Health and Safety Committee, in particular the Sampling and Analysis Subcommittee, BHSC Vice-Chair David Weitzman (US Department of Energy) and Gary Whitney (Los Alamos National Laboratory). Assistance from authors of the other chapters, and from Greg Day and Aleks Stafaniak at NIOSH, Morgantown, WV, is gratefully acknowledged. The efforts of many individuals toward establishment of a BeO reference material, most notably Larissa Welch and Tom Oatts (DOE Y-12 site), Greg Turk and Mike Winchester (NIST), Mark Hoover (NIOSH), and

Sam Johnson (US DOE), are also gratefully acknowledged. Finally, support from Savannah River Site personnel, including Maureen Bernard, Linda Youmans-McDonald, Steven Jahn, Marion Hook, and Edward Sadowski, has been invaluable.

References

1. C. Roelofs, *Preventing Hazards at the Source,* American Industrial Hygiene Association, Fairfax, VA, 2007.
2. WebElements Periodic Table of the Elements, www.webelements.com/beryllium, accessed 23 August 2008.
3. Spectrum Chemical Fact Sheet – Beryllium, www.speclab.com/elements/beryllium.htm, accessed 23 August 2008.
4. *Toxicological Profile for Beryllium*, US Agency for Toxic Substances and Disease Registry, Atlanta, GA, 2002.
5. M. E. Kolanz, *Appl. Occup. Environ. Hyg.*, 2001, **16**, 559–567.
6. Mineral Commodity Summaries, US Geological Survey, January 2008, www.usgs.gov/minerals/pubs/commodity/beryllium, accessed 25 August 2008.
7. O. A. Taiwo, Beryllium Sensitization in Aluminium Smelters, presented at the 3rd International Conference on Beryllium Disease, Philadelphia, PA, 2007, www.internationalbeconference07.com, accessed 25 August 2008.
8. O. A. Taiwo, M. D. Slade, L. F. Cantley, M. G. Fiellin, J. C. Wesdock, F. J. Bayer and M. R. Cullen, *J. Occup. Environ. Med.*, 2008, **50**, 157–162.
9. *Toxicological Profile for Beryllium*, US Agency for Toxic Substances and Disease Registry, Atlanta, GA, 2002.
10. J. E. Heber and T. B. Parsonage, in *Optical Materials and Structures Technologies, ed. W. A. Goodman,* Proceedings of the SPIE, 2003, 5179, 56–62.
11. Nickel Beryllium Alloys, Brush-Wellman, Mayfield Heights, OH, www.brushwellman.com, accessed 25 August 2008.
12. L. A. Maier, C. Gunn and L. S. Newman, *Beryllium Disease, in Environmental and Occupational Medicine,* ed. W. N. Rom and S. B. Markowitz, Wolters Kluwer/Lippincott Williams & Wilkins, Philadelphia, PA, 2006, pp. 1021–1037.
13. M. Eisenbud, C. F. Berghout and L. T. Steadman, *J. Ind. Hyg. Toxicol.*, 1948, **30**, 281–285.
14. J. H. Sterner and M. Eisenbud, *Arch. Ind. Hyg. Occup. Med.*, 1951, **4**, 123–151.
15. K. Kreiss, G. A. Day and C. R. Schuler, *Annu. Rev. Public Health*, 2007, **28**, 259–277.
16. P. C. Kelleher, J. W. Martyny, M. M. Mroz, L. A. Maier, A. J. Ruttenber, D. A. Young and L. S. Newman, *J. Occup. Environ. Med.*, 2001, **43**, 231–237.
17. A. K. Madl, K. Unice, J. L. Brown, M. E. Kolanz and M. S. Kent, *J. Occup. Environ. Hyg.*, 2007, **4**, 448–466.
18. K. Rosenman, V. Hertzberg, C. Rice, M. J. Reilly, J. Aronchick, J. E. Parker, J. Regovich and M. Rossman, *Environ. Health Perspect.*, 2005, **113**, 1366–1372.

19. A. P. Fontenot and L. A. Maier, *Trends Immunol.*, 2005, **26**, 543–549.
20. L. S. Newman, M. M. Mroz, R. Balkissoon and L. A. Maier, *Am. J. Respir. Crit. Care Med.*, 2005, **171**, 54–60.
21. G. H. Curtis, *AMA Arch. Derm. Syphilol.*, 1951, **64**, 470–482.
22. S. S. Tinkle, J. M. Antonini, B. A. Rich, J. R. Roberts and R. Salmen, *Environ. Health Perspect.*, 2003, **111**, 1202–1208.
23. G. A. Day, A. Dufresne, A. B. Stefaniak, C. R. Schuler, M. L. Stanton, W. E. Miller, M. S. Kent, D. C. Deubner, K. Kreiss and M. D. Hoover, *Ann. Occup. Hyg.*, 2007, **51**, 67–80.
24. *Summaries and Evaluations, Beryllium and Compounds*, Monograph Volume 58, International Agency for Research on Cancer, Lyon, France, 1994.
25. *Report on Carcinogens*, National Toxicology Program, Public Health Service, US Department of Health and Human Services, Research Triangle Park, NC, 11th edn, 2005.
26. M. E. Kolanz, A. K. Madl and M. A. Kelsh, *Appl. Occup. Environ. Hyg.*, 2001, **16**, 593–614.
27. P. S. Levy, H. D. Roth and D. C. Deubner, *J. Occup. Environ. Med.*, 2007, **49**, 96–101.
28. M. K. Schubauer-Berigan, J. A. Deddens, K. Steenland, W. T. Sanderson and M. R. Petersen, *Occup. Environ. Med.*, 2008, **65**, 379–383.
29. D. C. Deubner, P. S. Levy and H. D. Roth, Letter to the editor, *Occup. Environ. Med.*, 16 November 2007, http://oem.bmj.com, accessed 27 August 2008.
30. M. Eisenbud, *Environ. Res.*, 1982, **27**, 79–88.
31. US Code of Federal Regulations, 29 CFR Part 1910, Subpart Z: Toxic and Hazardous Substances, US Occupational Health and Safety Administration, Washington, DC.
32. US Code of Federal Regulations, 40 CFR Part 401.15, Toxic Pollutants, US Environmental Protection Agency, Washington, DC.
33. ACGIH® *Documentation for Beryllium and Compounds*, American Council of Governmental Industrial Hygienists, Cincinnati, OH, 2009.
34. M. Eisenbud, *Appl. Occup. Environ. Hyg.*, 1998, **13**, 25–31.
35. K. Kreiss, M. M. Mroz and L. S. Newman, *Am. J. Ind. Med.*, 1996, **30**, 16–25.
36. *TLVs® and BEIs®*, American Council of Governmental Industrial Hygienists, Cincinnati, OH, 1999, 2005, 2006, 2007, 2008 and 2009.
37. *Preliminary Initial Regulatory Flexibility Analysis of the Preliminary Draft Standard for Occupational Exposure to Beryllium*, US Occupational Safety and Health Administration, Washington DC, 2007.
38. US Code of Federal Regulations, 10 CFR Part 850, *Fed. Regist.*, 1999, 64(8th December), 68854–68914.
39. *Table AC-1 Permissible Exposure Limits for Chemical Contaminants*, California Occupational Health and Safety Standards Board, www.dir.ca.gov, accessed 29 August 2008.
40. *GESTIS International Limit Values for Chemical Agents*, Institut für Arbeitsschutz der Deutschen Gesetzlichen Unfallversicherung (BGIA), 2008, www.dguv.de/bgia/gestis-limit-values, accessed 27 August 2008.

41. National Research Council of the National Academies *Managing Health Effects of Beryllium Exposure*, National Academies Press, Washington, DC, 2008.

42. *Luckey Site, Luckey, OH. Record of Decision for Soils Operable Unit. Total Environmental Restoration Contract DACW27-7-D-0015*, US Army Corps of Engineers, 2006, www.lrb.usace.army.mil/fusrap/luckey/luckey-rod-soil-2006-07.pdf, accessed 29 August 2008.

43. A. Agrawal, J. P. Cronin, A. Agrawal, J. C. L. Tonazzi, L. Adams, K. Ashley, M. J. Brisson, B. Duran, G. Whitney, A. Burrell, T. M. McCleskey, J. Robbins and K. T. White, *Environ. Sci. Technol.*, 2008, **42**, 2066–2071.

44. *Annual Book of ASTM Standards, Volumes 11.01 and 11.02*, ASTM International, West Conshohocken, PA, updated annually.

45. *Methods for Chemical Analysis of Water and Wastes*, US Environmental Protection Agency, Washington, DC, 3rd edn, 1983.

46. *Standard Methods for the Examination of Water and Wastewater*, American Water Works Association, Denver, CO, 21st edn, 2005, www. standardmethods.org, accessed 30 August 2008.

47. M. Harper, *J. Environ. Monit.*, 2006, **8**, 598–604.

48. ISO 7708:1995, *Air Quality. Particle Size Fraction Definitions for Health Related Sampling*, International Organization for Standardization, Geneva, 1995.

49. *Environmental Health Criteria 106: Beryllium*, World Health Organization/ International Programme on Chemical Safety, Geneva, 1990, www. inchem.org/documents/ehc/ehc/ehc106.htm, accessed 29 August 2008.

50. J. S. Ignacio and W. H. Bullock, *A Strategy for Assessing and Managing Occupational Exposures,* American Industrial Hygiene Association, Fairfax, VA, 2006.

51. S. K. Dufay and M. M. Archuleta, *J. Environ. Monit.*, 2006, **8**, 630–633.

52. K. Kerr, *Sampling Beryllium Surface Contamination Using Wet, Dry, and Alcohol Wipe Methods*, Thesis submitted to Central Missouri State University, Warrensburg, MO, December 2004, www.osti.gov, accessed 15 September 2008.

53. CEN/TR 15278:2006, *Workplace Exposure. Strategy for the Evaluation of Dermal Exposure*, European Committee for Standardization, Brussels, 2006.

54. CEN/TS 15279:2006, Workplace Exposure. Measurement of Dermal Exposure. Principles and Methods, European Committee for Standardization, Brussels, 2006.

55. *Annual Book of ASTM Standards, Volumes 4.08 and 4.09*, ASTM International, West Conshohocken, PA, updated annually.

56. M. J. Brisson, K. Ashley, A. B. Stefaniak, A. A. Ekechukwu and K. L. Creek, *J. Environ. Monit.*, 2006, **8**, 605–611.

57. W. Pierce, S. M. Christian, M. J. Myers, J. D. Myers, Field-testing for environmental pollutants using briefcase sized portable LIBS system, presented at 3rd International Conference on Laser Induced Plasma Spectroscopy and Applications, Malaga, Spain, 2004.

58. Y. Su, Z. Jin, Y. Duan, M. Koby, V. Majidi, J. A. Olivares and S. P. Abeln, *Anal. Chim. Acta*, 2000, **422**, 209–216.
59. Y. Duan, Y. Su, Z. Jin and S. P. Abeln, *Anal. Chem.*, 2000, **72**, 1672–1679.
60. Researchers gain ground in air monitoring, in *Y-12 Report*, Summer 2004, www.y12.doe.gov/news/report/toc.php?vn=1_1, accessed 12 May 2006.
61. T. Vo-Dinh, M. Martin and B. Phifer, Real-Time Surface Contamination Monitor for Beryllium Oxide Particulates, presented at 2nd Symposium on Beryllium Particulates and Their Detection, Salt Lake City, UT, November 2005, www.sandia.gov/BHSC, accessed 12 May 2006.
62. M. J. Brisson, A. A. Ekechukwu, K. Ashley and S. D. Jahn, *J. ASTM Int.*, 2006, **3**, DOI 10:1039/b601524g.
63. A. B. Stefaniak, M. D. Hoover, G. A. Day, A. A. Ekechukwu, G. E. Whitney, C. A. Brink, R. C. Scripsick, *J. ASTM Int.*, 2005, **2**, DOI 10:1520/JAI13174.
64. A. B. Stefaniak, C. A. Brink, R. M. Dickerson, G. A. Day, M. J. Brisson, M. D. Hoover and R. C. Scripsick, *Anal. Bioanal. Chem.*, 2007, **387**, 2411–2417.
65. R. L. Watters Jr, M. D. Hoover, G. A. Day and A. B. Stefaniak, *J. ASTM Int.*, 2006, **2**, DOI 10:1520/JAI13171.
66. *Certificate of Analysis, Standard Reference Material 1877, Beryllium Oxide Powder*, National Institute of Standards and Technology, Gaithersburg, MD, US, May 20, 2008, www.nist.gov, accessed 30 August 2008.
67. Y. Thomassen, Occupational Beryllium Exposure in Primary Aluminium Production, presented at International Beryllium Research Conference, Montreal, QC, March 2005, www.irsst.qc.ca/files/documents/PublIRSST/Be-2005/Session-8/Thomassen.pdf, accessed 30 August 2008.

CHAPTER 2
Air Sampling[*][‡]

MARTIN HARPER

Chief, Exposure Assessment Branch, Health Effects Laboratory Division, National Institute for Occupational Safety and Health, 1095 Willowdale Rd., MS-3030, Morgantown, WV 26505, USA

Abstract

Substances that are airborne and toxic by inhalation are usually evaluated by sampling and analysis of the relative or absolute concentration of the substance in air. First it is necessary to state the reason for evaluation: determining the sources of emissions in work processes; determining the quantity of emissions to the environment; determining compliance with regulations and guidelines; determining the exposure distribution for affected persons; determining the effect of process modification on exposure; and determining risk of disease. Once the desired outcome has been clearly stated it is possible to consider a strategy for sampling that involves sample number, distribution and timing. The sampling strategy also affects the selection of sampling and analysis methods, including requirements of selectivity, sensitivity and accuracy. Methods may be prescriptive, and following such methods strictly may be a requirement under regulation or for accreditation. Conversely, today's

---

[*] This article was prepared by a US Government employee as part of his official duties. The US Government retains a nonexclusive, paid-up, irrevocable license to publish or reproduce this work, or allow others to do so for US Government purposes.

[‡] *Disclaimer*: The findings and conclusions in this report are those of the author and do not necessarily represent the views of the National Institute for Occupational Safety and Health. Mention of sampling devices does not constitute an endorsement, and does not imply that other devices are not fit for the same purpose. Discussion of the draft ACGIH® TLV® documentation does not imply acceptance or endorsement by the US Federal Government.

Beryllium: Environmental Analysis and Monitoring
Edited by Michael J. Brisson and Amy A. Ekechukwu
© Royal Society of Chemistry 2009
Published by the Royal Society of Chemistry, www.rsc.org

technology allows many options and variations that may not have been available in the past. Changing requirements, such as changes to concentration limit values, require a degree of flexibility. This chapter describes the theory and technology for sampling aerosols in general and beryllium aerosols in particular, with sufficient background and discussion to allow the user to select with confidence from the available options the tools and techniques most applicable to the desired outcome. It must be kept in mind that an air sample is only ever an estimate of exposure with an uncertainty that is associated with the methodology both of the strategy and the procedure.

2.1 Introduction

Air sampling is important because breathing a toxin is often the most important route of exposure. The high surface area of the lungs allows the absorption of gases and vapors and soluble airborne particles (aerosols), which can then exert their toxicity on some other organ. In addition, particles can also exert a toxic action on specific sites within the airways system. Some aerosols may have multiple sites of action. For example, inhalation of beryllium particles may result both in sensitization through absorption anywhere in the airways system and granulomas through a specific toxic action in the alveoli.[1] The human airways can be divided into different regions characterized by the probability of a particle entering that region. The probability is determined by the aerodynamic diameter (the diameter of the spherical particle of unit density with the same settling rate in still air as the particle of interest). The probability of deposition for large particles is effectively the same as the probability of penetration, but for particles in a smaller size range there may be a greater probability of being exhaled than deposited.[2] For particles in the nanometer size range, the likelihood of deposition rises as a result of particle diffusion, but the site of deposition may be in the nose or other parts of the upper airways. Deposition of particles in different regions usually leads to different physiological responses and the variation in response can be extreme. For example, a particle of certain size may cause damage, disease and even death, while a larger particle of the same composition may be relatively harmless. Thus the critical properties of a particle include the aerodynamic size as well as the potential biological or clinical reaction to its presence. Sampling of aerosols thus requires consideration of the size range(s) of interest.

   Aerosol monitoring can be either through the use of direct-reading instruments operating in near real-time or by capturing the aerosol constituents from a known volume of air for later analysis (sampling). No direct-reading instruments are considered sufficiently precise and accurate to be used for the purpose of compliance with regulated limit values, which are based on average concentrations over a defined period of time, although developments are trending towards that goal.[3] Thus aerosol sampling is the standard procedure for obtaining time-weighted average (TWA) concentrations for comparison with regulatory limit values or other guidelines. Several procedures exist for capturing particles from air and an excellent review has been published.[4] Analysis has typically been at an off-site laboratory, but field portable analytical methods may offer similar limits of detection and accuracy, with advantages of faster turnaround and, therefore, faster response to unhygienic conditions.

Air samples may be taken for several different purposes: to screen for the presence of an airborne contaminant; to determine the source of contaminant release; to determine the effect of a process change on exposure; to ensure compliance with a regulated or advisory limit value; or to determine the long-term exposure history of a worker or group of workers. Different sampling strategies are normally employed in service of these aims, even when the selection of sampling methodology is the same. Certain types of sampling methods may have specific advantages for certain strategies. For example, direct-reading instruments may be able to determine presence or absence, or the source or timing of a release, or the relative effect of process changes, but are rarely employed for accurate TWA measurements. The questions of who or where to sample, and for how long and how often, are important and the answers depend on the purpose(s) for which the information will be used. Sampling strategies are developed in response to specific needs.

Inhalation of contaminants is an important route of exposure, but not the only route. Ingestion and uptake through the skin are also important, and should be considered in any overall exposure assessment model. Since surface contamination is often the result of aerosol deposition, committees that deal with the assessment of aerosol exposure by inhalation sometimes include the assessment of surface deposition and dermal exposure in their remit.

2.2 Sampling Strategies

2.2.1 Sampling for Compliance with a Limit Value

Limit values for the concentration of airborne chemicals may be guidelines issued by a professional body such as the Threshold Limit Values (TLVs®) of the American Conference of Governmental Industrial Hygienists (ACGIH®),[5] or by a government body such as the Recommended Exposure Limits (RELs) of the National Institute for Occupational Safety and Health (NIOSH).[6] Limit values may also be set under regulation, such as Permissible Exposure Limits (PELs) by the US Occupational Safety and Health Administration (OSHA).[7] In the UK, limit values were set by the Health and Safety Commission, which has now merged with the Health and Safety Executive.[8] Regulated limit values for a number of countries may be found on the BGIA – Institut für Arbeitsschutz der Deutschen Gesetzlichen Unfallversicherung website (www.hvbg.de/e/bia/gestis/limit_values/index.html).

Limit values also may be values determined internally by a user or supplier, and these are often described as "in-house" limit values. A general term for all such limit values is occupational exposure limit (OEL). Very few OELs are set through a consensus of all interested parties, and the actual values may be contentious. A significant factor in all OELs is the uncertainty in deriving the value. An OEL that is difficult to assess, or costly to achieve routinely, might be called into question. Uncertainty may lead to differences of opinion in the exact level of the OEL, and opinion may change with advancing scientific knowledge. Users of OELs should be aware of the background data and methodology of developing a specific limit value. An important issue with an OEL is significance of exceeding it ("overexposure"). A single instance of overexposure is normally sufficient to prove non-compliance with OELs prescribed by regulation. However, given the log-normal distributions (positively skewed on the normal scale) associated with exposure data, an occasional value exceeding the limit is reasonably common and more likely to occur as the number of measurements increases.[9] This acts as a disincentive to collecting more samples than absolutely necessary to demonstrate compliance. However, if it is assumed that OELs include safety factors and that most OELs are set to control long-term exposures to chronic toxins for which the long-term average is the best predictor of disease, then some rate of exceeding the OEL may be tolerable.[10] It not a simple matter to find agreement on what the maximum allowable rate and magnitude of such overexposures might be. It may not be possible to reconcile using an OEL for risk management as well as risk assessment.[11]

The typical strategy for determining compliance with a limit value is to take a single sample from the worker judged to be at greatest risk for overexposure. This strategy often is ascribed to OSHA, although the OSHA technical manual actually suggests the compliance officer should "take a sufficient number of samples to obtain a representative estimate of exposure[12]. OSHA compliance officers are unlikely to have the time or resources to take many samples. Thus it is also recommended to "consult the *NIOSH Occupational Exposure Sampling Strategy Manual* for further information." This NIOSH manual[13] includes a technique termed the Employee Exposure Determination and Measurement Strategy, which involves sampling the worst-case exposure with the philosophy that, by selecting a worst-case exposure situation, it would rule out the need to do further sampling if this exposure is acceptable. However, even for expert industrial hygienists, it is probably not always possible to correctly identify the individual with the highest exposure, and the chance of the compliance officer actually being present at the time the highest exposure is occurring is remote. Thus, while a single measurement exceeding the OEL proves non-compliance, a single measurement below does not prove compliance for all workers on all days. Based on assumptions about the distribution of exposures between workers considered to be in a similar exposure situation, the NIOSH manual suggests that, if the exposure concentration of the selected individual is less than half the limit value, there is a 95% probability that other workers exposures will not exceed this standard. This is the concept behind the "action level" at which it might be presumed prudent to implement some form of preventive action (enhanced frequency of monitoring, additional engineering controls, personal protective equipment, medical monitoring, *etc.*). The action level takes into account between-worker variation, but does not necessarily take into account the day-to-day variation in exposures which may exceed an order of magnitude. Therefore, the population of potentially overexposed workers might be underestimated. Re-sampling individuals on subsequent days gives greater confidence in the safety of the environment. The European standard EN 689:1996, for example, includes strategies for the frequency of re-sampling individuals based on the magnitude of successive results in comparison to the OEL.[14] Guidelines generally mandate re-sampling in the event of any changes in process, equipment or task. The action level's 95% probability that exposures will not exceed the OEL also implies that 5% of samples will exceed the OEL. Whether this magnitude of overexposure should be tolerated is open to question, and its significance may differ for acute and chronic toxins. Rappaport *et al.* recommended a minimum of 20 samples (two each from 10 workers) with a group mean less than one-fifth the OEL as a determinant (90% probability) of whether any worker's overall mean exposure will exceed the OEL.[15] The implication of an overall mean exposure is that the result could exceed the OEL up to 50% of the time. Rappaport has argued that OELs for chronic toxins should be regarded as a long-term mean, rather than a daily maximum,[9] although others consider this less protective.[11]

For substances where exposure can result in chronic health effects, overexposures are generally better tolerated than are overexposures to acute toxins. Characterizing the upper end of the exposure distribution precisely requires substantial resources, but these exposures contribute most towards acutely toxic effects, and thus there is justification for this exercise.[16] Toxins with an immediate acute effect, and toxins where short episodes of high exposures can precipitate an adverse long-term effect, are often controlled through a Short Term Exposure Limit or STEL. Where both a long-term OEL and a STEL apply, the STEL is higher but averaged over a much shorter period, typically 15 minutes. STELs set by ACGIH® also include important limits on the number of times per shift the TLV® can be exceeded during episodes where the STEL might be exceeded, and on the minimum length of time that should elapse between such episodes. Compliance with all these requirements of a STEL greatly improves the likelihood that the 8-hour TLV® is also not exceeded.[17] Full-shift limit values typically are based on an 8-hour shift and a 40-hour work week. Strategies for adjusting OELs for other working arrangements have been published.[18–21]

2.2.2 Sampling to Identify a Group Range of Exposures

The response of individual workers to similar exposures is highly variable. For epidemiological studies to provide meaningful results, large groups of workers are required with well-categorized exposures. One strategy for determining the exposures of a group of workers would be to sample them all, on every workday and, if resources allow, this clearly would be the best option. Usually resources do not allow, but it may be possible to sample just a few members of the group, provided it can be considered a homogenous exposure group (HEG), also referred to a similarly exposed group (SEG). The determinants of an HEG or SEG are usually process, job, task, or agent.[16] Exposure measurements made on the group can confirm the judgment used to define the group. It should be noted that HEGs are not common. In a study of 20 000 chemical exposures that focused only on workers with multiple measurements, it was found that only one-quarter of the groups based on job title and factory had 95% of individual mean exposures within a two-fold range.[22] Where a HEG can be established, it is possible to define a group mean exposure for epidemiological studies and to predict the likelihood of non-compliance with OELs. When the average exposure value is very much less than the OEL, but still detectable, a minimum of six samples may be enough to establish a 95% upper confidence limit on the mean also being less than the OEL.[16] A more refined strategy is required when the average exposure value approaches the OEL. It becomes necessary to measure the variance around the mean, for which a minimum of 11 samples is required. Models have been produced to deduce the sample size required as a function of geometric mean as a fraction of the OEL, and geometric standard deviation. As the mean approaches the OEL and as the standard deviation increases, the number of samples required to make a statistical prediction of compliance rises rapidly – into the hundreds. However, the incremental information added by the later samples decreases in relative value, while the costs continue to increase. As a cost-benefit trade-off, the number of samples to be taken may need to be curtailed. If all samples are taken in a quick campaign, the results may not necessarily reflect day-to-day variation. The best strategy for assigning an average exposure over a working lifetime is to spread the samples over randomly selected days of the year, during a period of several years. Rappaport *et al.*[15] suggest that increasing the number of measurements to more than four per worker is less valuable in determining the group mean than increasing the number of workers monitored per group (preferably > 10). For toxins that pose an acute risk, a one-sided tolerance interval (at 90% confidence) for the 95% percentile of the exposure distribution is recommended, with the upper tolerance limit being less than the OEL (STEL).[16]

The strategy detailed above is suitable for datasets that fit a parametric model. Non-parametric datasets require many more samples for characterization and are not well suited for statistical tests. Bimodal distributions are non-parametric. Their presence may be deduced from high geometric standard deviations.

Sampling results should also be interpreted through the eyes of professional judgment, whether or not there are sufficient for statistical analysis. Professional judgment can also be used in a Bayesian approach to refining scarce data.[23] As noted previously, there is a disincentive for employers to want to take any more samples than the minimum required. Thus a typical investigation may consist of a sampling campaign providing just a single result. This is a poor legacy to leave for future epidemiological researchers, or those charged with attempting historical exposure reconstruction.

2.2.3 Real-Time Monitoring

Real-time or, more accurately, near real-time (NRT) monitoring (monitor derives from the Latin: *monere* to warn) is the goal of much research in exposure assessment because of the potential advantages. In addition to being able to track rapid changes and determine the frequency, duration and size of peak exposures, manipulations of the collected data can also provide short- or long-term average concentrations, which are the basis of most OELs. Further, such instruments could provide information of immediate value to the workers such as high-concentration alarms or warnings regarding the direction of trends in exposure.

Personal real-time monitors exist for aerosols, but the measurements typically depend on properties that are characteristic of an aerosol of any composition and then cannot distinguish individual components of the aerosol.[24]

The simplest aerosol monitors measure the attenuation of light. Such opacity monitors are used to measure stack emissions, but they are not sufficiently sensitive for workplace aerosol measurements.[25] The most common aerosol monitors are light-scattering instruments (nephelometers) because these are sensitive, portable, relatively rugged and inexpensive.[26] Particles as small as 0.3 μm can be detected with incandescent light sources, and this can be lowered to 0.05 μm with laser illumination. Response is a function of particle size, becoming weaker as size increases. The upper size range for particle detection is between 10 and 20 μm. Unfortunately, these instruments detect aerosols based on their size, shape and refractive index, and not their mass, and, as noted, not on the amount of specific chemical such as beryllium they might contain. Because the physical and chemical characteristics of aerosols can change over time, it is not always possible to calibrate the response against the mass of a specific component. Thus light-scattering instruments are not usually used to demonstrate compliance with OELs.[27]

The factory calibration of a light-scattering instrument is typically to some standard material such as Arizona road dust. This calibration is usually quite different from a calibration based on the dust of interest. However, it is possible to calibrate response to the mass collected in the field, and the results are frequently linear.[28] When light-scattering instruments are used in the active mode (air is pulled through the sensor), it may be possible to back this up with a filter to collect the dust for subsequent weighing. When used in the passive mode (dust is allowed to pass into the sensor by air currents, *etc.*), an additional sampler must be collocated.

When particles are fed through a light beam individually, they can be counted, and instruments for this purpose are known as optical particle counters (OPC). Some can also give an indication of particle size. More sophisticated and expensive particle counters include the aerodynamic particle sizers (APS). Time-of-flight aerodynamic sizing accounts for particle shape and is unaffected by index of refraction or Mie scattering. Additional instruments to determine the number and count size distribution of ultrafine aerosols are described in Section 2.3.8 and in Pui and Chen (2005).[29]

Real-time monitoring of mass is possible with a micro-balance, such as an oscillating quartz crystal[30] (*e.g.* Air Particle Analyzer, California Measurements, Inc., Sierra Madre, CA) or vibrating tapered element[31] (*e.g.* Ambient Particle Monitor, Rupprecht & Patashnik Co., Inc., Albany, NY), where the frequency of oscillation or vibration changes as mass is applied. Beta-attenuation[32] is another methodology that can be used as a surrogate measure of mass (*e.g.* Anderson Instruments, Inc., Smyrna, GA). Most commercial instruments are large and developed for use in fixed-site ambient air monitoring. A personal tapered-element oscillating microbalance (TEOM) has recently been developed for real-time respirable mass measurement in coal mines.[33] None of these instruments are specific to a single element within the mass.

An early method for the continuous quantitative determination of beryllium in air is described in 1965.[34] Based on an even earlier semi-quantitative procedure, it features collection of aerosol on a filter paper tape for a defined period of time, followed by introduction of the sample into a spark gap between platinum-capped copper electrodes and detection of the signal through a spectrometer. It was considered to be accurate to within about 20% down to concentrations of $0.2\,\mu g\,m^{-3}$, with detection possible at $0.05\,\mu g\,m^{-3}$ for a 75 second cycle. It had an alarm feature if concentrations exceeded the then "toxic limit" of $2\,\mu g\,m^{-3}$.

Two more recently developed instruments that can measure beryllium aerosols are the aerosol time-of-flight mass spectrometer (ATOFMS),[35] which with aerosol beam focusing has been reported to have a detection limit for beryllium of $0.006\,\mu g\,m^{-3}$, and the aerosol beam-focused laser-induced plasma spectrometer (ABFLIPS),[36] which with 30 minute averaging has been shown to have a detection limit of $0.15\,\mu g\,m^{-3}$ for in-stack particulate.[35] ABFLIPS uses focusing of a high-energy neodymium/yttrium–aluminium–garnet (Nd/YAG) laser to produce microplasma, which vaporizes any particles and causes atomic emission. Both of these instruments are very large and expensive at the present time.

A prototype microwave-induced plasma torch for atomic emission spectroscopy (MIPAES) has been tested on solutions of beryllium, but it is not known if it has yet reached the stage of an aerosol monitor.[37] A potentially novel approach that has been studied is a portable, capillary electrophoresis microchip with sampling from air *via* a microimpinger, which is integrated onto the microchip itself, and detection using metallochromic dyes and fluorophores in combination with small, inexpensive light sources (*e.g.* LEDs) and photodetectors.[38]

2.2.4   Area *Versus* Personal Sampling

In general, where personnel exposures are to be measured, normal practice is to place the sampler on the person, preferably within their "breathing zone", *i.e.* the volume around the person from which the breath is drawn. Several definitions of breathing zone exist; a typical one being a hemisphere of 30 cm centered on the mid-point between the mouth and nose. Little work exists to validate how concentrations may vary across such a space. Tests on radioactively labeled ambient aerosol particles (activity median diameter=0.15 μm) and dusts with mass median aerodynamic diameters of 1.6 and 8.0 μm concluded that no bias results from location for aerosols dispersed uniformly within the breathing zone,[39] and some more recent work on aerosols in the coal mine environment suggests that spatial variability and instrument precision have a greater effect on measurement than sampler location within the breathing zone.[40] This may not also be the case for coarse aerosols generated near to the worker's waist level. The presence of clothing such as aprons or billed caps, as well as the local heating from the body, may affect air currents around a worker leading to differences between personal and area sampling results.[41] The type of clothing may have an effect on sample collection through immediate particle bounce or particle capture and later re-entrainment.[42] However, these bounced particles may also be available for inhalation.

Personal sampling requires the cooperation of the worker, who has to carry the sampling apparatus around while performing his or her duties. This cooperation may not be easy to obtain in jobs that involve arduous work, excessive heat, or confined spaces. Pump size and weight are the critical issues here, and they necessarily increase with increasing flow-rate capability. Higher flow-rates induce greater pressure drop across a filter, which causes greater drain on battery power. The pressure drop across a filter is also a function of the area of the filter and its porosity. Thus higher flow-rates can be maintained for longer when the filter area is larger and the porosity is greater (see filters, Section 2.4). Battery technology is also improving. Lithium-ion and nickel–metal hydride batteries with higher power-density are replacing nickel–cadmium batteries. However, all these types of batteries require recharge and this can be a tedious and demanding task. Pumps are now available that will run on dry cells, which are much more convenient when large numbers of pumps are required on a daily basis or when the need for a rapid response is anticipated.

When measurements are being taken to assess workers' exposures, it is good practice and it may be required under law to provide the results to the worker. Sometimes measurements are not taken on the worker, for example when it is desired to determine the concentration distribution around a task, or in a building, or where the need is to determine the sources of contaminant leaks. In this case, fixed-point or "area" measurements are made, usually with the samplers located according to a pre-determined plan. It may also be necessary to take area measurements for worker exposure assessment where it is impractical or undesirable for a worker to carry the sampling equipment. In such cases, the samplers are located as close as possible to the worker's breathing zone, and may need to be moved in response to changes in the worker's position.

A particular case relevant to the assessment of beryllium exposures is the need on occasion to take "high-volume" samples, usually to address poor analytical sensitivity. The pumps required for such samples are often too large to be carried by the worker. An alternative strategy in this instance is to take a personal sample with a single sampler over multiple shifts, being careful to preserve the collected material between sampling periods. This allows a higher volume of air to be sampled, but at the expense of averaging the result over several work shifts.

2.2.5 Choice of Sampling Time

Different time periods of monitoring may be required depending on factors including the nature of the potential injury resulting from overexposure and considerations of control. For many chemicals encountered in the workplace, time-weighted averages based on a work shift, typically of 8-hours duration, repeated 5 days in a week, are the best predictors of risk. For other chemicals, particularly those causing acute responses, such as sensitization, it may be the magnitude and number of excursion events. Monitoring short-term events may also be an integral part of the exposure control strategy. However, for beryllium, sensitivity is definitely an issue. As discussed in Chapter 1, for a proposed STEL of $0.2\,\mu g\,m^{-3}$ and a sample pump flow rate of $2\,L\,min^{-1}$, a sensitivity of 0.6 ng would be required. By comparison, the NIOSH methods for graphite furnace atomic absorption and field-portable fluorimetry have limits of quantitation of 50 and 5 ng, respectively.[43,44] There could easily be more than one 15-minute period of exposures of concern during a work shift and thus a strategy of monitoring short-term events is likely to prove more costly than one based on a time-weighted full shift average.

2.3 Aerosols

Aerosols are particles, solid or liquid, suspended in air. They may have momentum imparted to them by the process of formation, they may be buffeted by air movements, and they fall under the influence of gravity. Particle shape, density, and size all play important roles in how these processes affect an aerosol. For example, larger and heavier particles have greater inertia but tend to settle faster than smaller and less dense particles. A great deal of particle behavior can be predicted from its aerodynamic equivalent diameter (AED), which is the diameter of an equivalent spherical particle of unit density having the same settling velocity. Particles with similar AED will behave similarly, *i.e.* they will have equal probability of penetration to the same region of the human airways and will have equal probability of deposition. A selection of useful resources on aerosol behavior is provided at the end of the chapter.

In order to understand the risks posed by aerosols, it is first necessary to discuss their behavior in relation to human physiology. The primary risk from an aerosol is that it can be breathed into the airways where particles may deposit.[45] As air is breathed in it passes through a tortuous path and its velocity is reduced. Large particles may deposit by impaction where direction changes, and by gravitational settling with reduced air velocity. On the other hand, very small particles may diffuse to the walls of the airways by Brownian motion. In between, there is a region of relatively small particles whose probability of deposition may be quite low. Such particles may be breathed out as well as in. However, because it is difficult to develop sampling devices that exclude particles in this region, the ability of a particle to penetrate the airways system is currently used as the basis for measuring exposures. It is possible that the probability of deposition, which is the more accurate measure of risk, might be used in the future.[2] Recognition that disease can result from where a particle deposits in addition to its nature has led to the development of size ranges of concern for different aerosols, although these size ranges are based on the inhalation characteristics of a "reference worker" from at least 50 years ago[46] whose age, gender, and physiognomy might be very different today.[47] The metric of aerosol exposure is currently mass, although particle number has been used in the past. Mass is certainly appropriate for materials that result in systemic toxicity as is the case, for example, with pesticides and many metals. However, for other outcomes, particle number or surface area might be a more appropriate metric.

2.3.1 Sources and Types of Beryllium Aerosols

Beryllium is mined as beryl or bertrandite $[Be_4Si_2O_7(OH)_2]$.[48] These silicates are extremely insoluble and there appears to be little if any disease associated with mining, even though mining traditionally involves the production of large quantities of airborne dust.[49] The ores are digested in a wet process whose end product is beryllium hydroxide.[50] This can be further processed into beryllium oxide, beryllium metal, and beryllium alloys with other metals, most notably copper.

Beryllium oxide is produced by calcining the hydroxide. The screened powder is typically cold-pressed into solid pieces, which are machined and polished. Aerosols are likely to be produced at the screening stage, and may also occur in the machining and polishing stages, especially if this is not a wet process. Production of pure beryllium metal involves converting the hydroxide to the volatile fluoride and reducing it to produce powdered metal. Beryllium fluoride vapor may give rise to condensation aerosols, and handling the beryllium metal powder may give rise to an aerosol.

Beryllium is alloyed with other metals in processes that typically involve melting the metal in a furnace. Furnace operations are traditionally dusty. Once beryllium products are produced, they may undergo secondary casting, machining or polishing for specific applications, and all of these processes can lead to aerosol formation.[50,51]

Beryllium is also found as a contaminant of the aluminium ore, bauxite, with the content varying from 0.01–4.0 ppm with geographic source. The treatment of bauxite to form the alumina feed for aluminium production does not remove beryllium, which can react with the fluorides in the aluminium reduction process to produce beryllium fluoride whose sublimation temperature (800 °C) is less than the temperature of the bath (960 °C). Personal inhalable samples ($n=274$) from workers in Norwegian aluminium potrooms found beryllium concentrations ranging from $0.013 \, \mu g \, m^{-3}$ to $0.34 \, \mu g \, m^{-3}$ with a geometric mean of $0.02 \, \mu g \, m^{-3}$.[52] About a quarter of the inhalable sample was in the respirable particle size fraction and around 80% of the total was found to be water-soluble. In another study of four smelters in Canada, the US and Italy, the geometric mean exposure from all four sites combined was $0.05 \, \mu g \, m^{-3}$, with a high value of $13 \, \mu g \, m^{-3}$. Two workers were reported to be sensitized.[53]

Interestingly, beryllium is often added back to aluminium, either to make specific low density alloys or materials, or to prevent spinel oxide formation in high magnesium alloys or to improve the casting of aluminium. In addition, scrap aluminium, potentially containing beryllium, is often also used in casting. While beryllium could not be detected in a casting house in a recent study,[54] the limit of detection for the short-period low flow-rate samples collected (100 minutes at $2 \, L \, min^{-1}$) can be estimated as a minimum of $1 \, \mu g \, m^{-3}$, and thus the hazard was not well quantified.

Welding and machining of these products could produce further aerosols. A study of beryllium emissions from cast aluminium welding concluded that as little as 0.0008% beryllium in the alloy could lead to concentrations of beryllium in the welding fume exceeding $2 \, \mu g \, m^{-3}$.[55] Beryllium is also found naturally in coal, and power station emissions may contain particulate beryllium.[56] Slag made from coal used for abrasive blasting may also produce an aerosol containing beryllium.[57] Finally, there is the disposal or recycling of spent beryllium products,[58] which may also be a source of aerosol.[59]

2.3.2 Aerosol Sampling

In the early part of the 20th century, particle concentrations were measured in terms of number, as, for example, millions of particles per cubic foot of air (mppcf). At one time, particles were collected by pulling a known volume of air through a tube filled with sugar.[60] The contents could then be emptied and the sugar dissolved in a known quantity of liquid. The resulting suspension of insoluble dust could be analyzed under a microscope, by sizing and counting. The impinger sampler was an improvement whereby the aerosol was collected directly into a liquid.[61] Particles of all sizes below a certain limit (*e.g.* 5 or 10 μm projected area diameter) were counted equally. The impinger was used with a vacuum pump, typically for area sampling. If personal sampling was preferred, the hygienist would have to maintain the sampler as close as possible to the worker, a very cumbersome arrangement. The personal sampling pump was designed in the 1960s[62] and the impinger was miniaturized[63] so that personal samples could be taken more easily. Mass is the preferred metric for chemicals with a systemic toxic effect, as is the case with many metals, and the ability to measure small masses of chemicals amidst larger quantities of background dust was provided by advances in analytical instrumentation in the 1950s, particularly atomic absorption spectroscopy for metals.

The advantages of filters (no liquid in glass) for chemical analysis were apparent.[64] In 1944, the US Bureau of Mines compared filter sampling to the impinger and, in 1957, the US Public Health Service (now NIOSH) revealed it had been using membrane filters for several years.[65] Filters require a holder to which vacuum can be applied. Early holders were simply supports that exposed the filter to the air, which had the drawback of leaving them free to be damaged or tampered with. The plastic closed-face cassette (CFC), which houses 37 mm diameter filters, then known as the "Millipore Monitor", was developed in 1956 for "clean–room" applications[66] and it appears in 1960 in the first edition of the American Conference of Governmental Industrial Hygienists (ACGIH®) *Air Sampling Instruments Handbook*.[67] The small (4 mm) air entry inlet of the CFC makes the filter much less likely to be accidentally or deliberately damaged, and thus it became very popular. NIOSH Method S349[68] for total suspended particulate (TSP), sometimes known as "total dust", using the CFC was validated and published in 1977, and further tests were carried out to investigate performance in the early 1980s.[69–71]

Throughout earlier periods, the main focus had been on particles smaller than about 10 μm AED, which are sampled relatively efficiently by the CFC. By the mid-1980s, it became clear that the CFC sampled larger aerosols very poorly, based on tests that only included the particulate mass collected on the filter. When aerosol is collected on filters without size selection and analyzed gravimetrically or chemically, the cubic relationship between volume (related to mass) and diameter means large particles assumes much greater importance than when the metric is particle number.

2.3.3 Size-selective Sampling

Different particle sizes will reach different regions of the human airways. For some particles which are readily soluble and cause systemic toxicity, there may be no difference in consequence of where they deposit. For particles which have a specific site of interaction in the airways, size may be crucial in determining the hazard. For example, some particles have a site of action in the nose (*e.g.* causing sino-nasal cancer), the pharynx (*e.g.* pharyngeal cancer), the upper airways (*e.g.* causing bronchoconstriction), or the alveolar region (*e.g.* causing pneumoconiosis). For such particles, it is useful to define size fractions of concern. Much work has gone into defining and validating the human aspiration efficiency for aerosols, but there are many variables involved, in particular the rate and depth of breathing, the ratio of nose to mouth breathing (all of which may vary with age, gender, physiognomy, and the health status of the individual), and the local air movement (speed and direction). In general, smaller particles are inhaled with a high efficiency, and this efficiency falls off for larger particles. There is an upper limit beyond which the probability of inhalation is virtually zero, but this point has not yet been agreed by researchers, in part because of the difficulty of experimentation with such large particles.[47] Appropriate size fractions of concern can be incorporated into OELs. The ACGIH® Threshold Limit Value (TLV®) committee[72] accepted the three fractions agreed by international consensus (ISO 7708),[73] and began to examine aerosol TLVs® to determine which fraction was most applicable to each. Even though this process had been taking place since 1993, five years later an industrial hygienist (IH) in a letter to the editor of the American Industrial Hygiene Association Journal stated: "*This is such a fundamental change . . . in particle evaluation . . . I'm a practicing IH . . . I knew there was some research . . . but I didn't realize it had developed to this point . . . I should have been following this issue closely*". The author felt "*like I've had someone slap me across the face.*[74]

2.3.4 The Inhalable Convention

In the region where experiments are relatively easy to perform, *i.e.* with particles up to an AED of 100 μm, the performance characteristics for a sampler intended to mimic breathing through the nose and mouth has been agreed, at least for wind speeds exceeding $1 \, \mathrm{m \, sec^{-1}}$, and this has been dubbed the "inhalable" convention.[75–78] The shape of the efficiency curve is unusual in that it falls off between 100% for particles of $1 \, \mu m$ AED to 50% at $50 \, \mu m$ AED, but then is constant between 50 and $100 \, \mu m$ AED. However, it has been shown that, in many workplaces, local wind speeds around the worker do not exceed $1 \, \mathrm{m \, sec^{-1}}$ [79] and some work has been carried out to validate the most appropriate aspiration efficiency under slowly moving air (sometimes referred to as "calm air") conditions.[80–82]

Once a convention for the aspiration efficiency of the human head had been defined, particle size-selective samplers could then be developed which would match that efficiency and thus provide a more useful estimate of dose. Nonetheless, only one sampler, the UK Institute of Occupational Medicine (IOM) sampler, was developed specifically to meet the inhalable convention.[83] Many samplers designed before the concept of inhalable sampling was introduced remain in routine use. The earliest were simply holders for a filter; later, devices were designed to protect the filter from disturbance or damage, and, in some cases, to have a specific efficiency of aerosol aspiration. A protocol exists to allow samplers to be tested against the inhalable convention,[84] and this has been done, showing that some existing samplers perform quite closely to the inhalable convention, but others less so.[85] Whatever has been the performance of a sampler in laboratory tests, it is interesting to note that few have been completely abandoned in general use; this is because the sampling device may be stipulated in regulation, or there has been a desire to maintain continuity of methodology with that used in the past, or the sampler has some other advantageous design characteristic, such as ease of use or calibration, or low cost and disposability. At least one pre-existing sampler has been re-designed to more closely match the inhalable convention.

Maintaining continuity with historical methods is often used as an argument against change. However the time-frame of continuity is limited by changes that have been made in sampling methodology over time. For example, in the US, aerosol sampling methodology has evolved from a tube containing sugar, through impinger to midget impinger to filter cassette over the past 100 years, and these transitions have already had consequences for the setting of target values and for epidemiological studies.

The ACGIH® has been examining its threshold limit values that have traditionally been referred to TSP to determine whether the inhalable sampling convention should apply to these substances. Larger particles are removed from the airways through the mucociliary escalator, and can be expectorated, swallowed, and absorbed through the gut. Therefore, the inhalable convention has been deemed appropriate for substances where the total mass dose is considered important, for example, in the case of organophosphorous pesticides. Many metals and their compounds, including nickel and molybdenum, also have received inhalable designations, and in 2009, ACGIH® applied the inhalable designation to beryllium.[5] The inhalable convention represents the largest particle-size portion of aerosol that can be inhaled, and in the case of beryllium, it is thought that the maximum potential dose should be assessed in order to prevent sensitization. However, the inhalable fraction may not be the only fraction of interest, especially for the progression from sensitization to *chronic beryllium disease* (CBD). A different convention, *e.g.* the respirable convention, may turn out to be an appropriate selection.[86]

One issue with specifying an OEL with respect to the inhalable convention is the effect on industries where previously TSP sampling had taken place. Side-by-side comparisons between filter samples from the CFC and samples collected with samplers designed for the inhalable convention have shown significant bias depending on the size distribution of the aerosol, with coarser dusts showing the greatest difference.[87] Since the median difference is a factor of 2.5 and published data include individual pair differences up to a factor of 70, it is clear that there is an increased likelihood of obtaining a value exceeding the OEL with an inhalable sampler, even if the OEL is adjusted for the median difference. It appeared in actual practice that many TLVs® were being adjusted by simply applying an inhalable specification without change to the adopted value.[88] This may be justified if it is known that the particles are generally small, so that there would be little difference in collection between an inhalable or other sampler. If the TLV® was set based on animal experiments where particle size was not controlled, this adjustment may also be valid. However, if the basis of the OEL was epidemiology derived from workplace samples using TSP samplers, such as the CFC, then the adjustment could be questionable. In the case of beryllium, many past workplace samples have been taken using so-called "high-volume" samplers whose aspiration efficiency is unknown, but known to be different from the CFC. It may be prudent to consider that the collection efficiency might have been greater than the CFC for large particles and thus might have been more like the inhalable convention. The ACGIH® has responded to the general issues of converting TLVs® from "TSP" to "inhalable" in more detail.[89] However, anomalies may still occur. For example, it was pointed out that if the coarse dust adjustment was used on the soapstone TLV® (which was converted to "inhalable" without change in value), the result obtained would be less than the respirable TLV® – a nonsensical situation.[90]

2.3.5 Thoracic Convention

The thoracic convention is a subset of the inhalable convention for particles capable of passing the head airways. Because the 50% probability of penetration (D_{50}) for this convention is found at $10\,\mu m$ AED, it is often equated with the US Environmental Protection Agency (EPA) PM10 standard (particulate matter $<10\,\mu m$ AED). However, the PM10 convention has a much sharper drop in penetration with particle size than the thoracic convention (which makes it more convenient to sample with sharp-cut impactors).

Few substances have so far received a notation to be sampled according to the thoracic convention. NIOSH has a recommended exposure limit (REL) for metalworking fluids[91] which suggests using thoracic samplers (when available), and the ACGIH® has included sulfuric acid mist on its notice of intended change to be sampled according to the thoracic convention.[92]

It has been suggested that the poor efficiency by which larger particles reach the filter of the CFC means the filter-catch of the CFC may be closer to a thoracic sample. However, this is not entirely accurate. An initial comparison between the CFC filter and a recently developed thoracic cyclone suggested the need for a 25% correction factor in the case of metalworking fluids,[93] while more recent work suggests this correction factor might not be constant across all industries.[94]

2.3.6 Respirable Conventions

The concept of separately assessing fine particulate that can penetrate to the alveolar region predates the research into general inhalability, and stems from the high prevalence of pneumoconiosis, especially silicosis, among miners and stone-cutters. The fraction of fine dust that can penetrate so deeply into the lungs is known as the respirable fraction. There have been suggestions that assessment of the respirable fraction of beryllium might be important on account of the granulomas associated with CBD.[95]

Several respirable conventions have been widely used. The three most important are those which were put forward: (a) by the British Medical Research Council, often referred to as the Johannesburg convention;[96] (b) by a meeting in January 1961 sponsored by the US Atomic Energy Commission, Office of Health and Safety, which was later adopted by the ACGIH® and by OSHA;[97] and (c) by the International Organization for Standardization (ISO),[73] which is currently accepted by ACGIH® and NIOSH.

All three conventions suggest similar high probabilities of penetration for small particles <1 μm AED) and low probabilities of penetration for large particles >10 μm AED). However, they differ in between these values. Respirable conventions have often been referred to by the size at which the probability of penetration is 50% (although the entire probability curve is different for the different respirable conventions). For the three conventions listed above, the 50% probability of penetration (D_{50}) is found at 5.0, 3.5 and 4.0 μm AED, respectively.[98]

The ISO curve is to some extent a compromise between the other two.[73] Size-selective conventions that match the ISO definitions were incorporated by reference in workplace standards of the European Union (EU), but the status of these standards is now "practical guidelines of a non-binding nature", allowing EU countries to choose their own definitions and methods.[99] Therefore ISO conventions are used in many, but not all, EU countries.

2.3.7 High Volume Sampling

Where it is necessary to assess very low exposures, it may be necessary to collect samples from a large volume of air to meet limitations of analytical sensitivity. Beryllium is a material for which some of the OELs are very low. In addition, in recognition of the potential hazard, many beryllium workplaces are kept very clean, so that the intention in sampling these environments is to measure even lower concentrations. Where it is not possible to enhance analytical capabilities further (at least in the short term), it may be possible to increase sensitivity by increasing the size of the sample. However, pulling more air through a filter can also have its drawbacks.

In order to pull more air through a filter one can increase sampling time or flow-rate. Increasing sampling time may be impossible if it is necessary to demonstrate compliance with a specific time-weighted average exposure period. For example, while it may be possible to keep re-using the same sampler over successive shifts until sufficient material has been collected for analysis,[100] it would not be possible to tell whether an overexposure occurred on one shift but was then balanced by compliant exposures on others. Even if the results are so high that it is likely that a full shift overexposure took place, it would still not be possible to determine on which day or days it took place. Continued re-use of the same sampler also increases the risk of sample loss. Increasing the flow-rate through the sampler may require heavy, mains-operated vacuum pumps to deal with the higher pressure drop across the filter, although it may be possible to substitute filters with larger pore-sizes to compensate. Increasing the flow-rate through a size-selective sampler changes the specific size-selection character-istics. Finally, the limiting situation is when the high-volume sampler begins to re-entrain its own, cleaned exhaust.

Personal high-volume pumps are available in the 8–15 L min^{-1} range, but they may require open-pore filters, such as 8 μm mixed cellulose ester (MCE) or glass-fiber filters. Vacuum pumps can operate over a similar range with standard 0.8 μm MCE filters, but these are not personal pumps. Collecting large samples of air for very low concentrations of specific substances usually also collects large quantities of other substances which might interfere with the analysis.

2.3.8 Ultra-fine Particle Sampling

The definitions of ultra-fine or nanoparticles size ranges are not presently fixed. Often, ultra-fine refers to particles below 1 μm and nanoparticles to those less than 100 nm diameter. However, there is considerable discussion.[101] In this region, mass is not a particularly useful metric, and it is often suggested that it be replaced by particle number or surface area.

Particles of these dimensions are the result of condensation from the vapor phase, perhaps enhanced by nucleation and agglomeration. Thus, they are typically encountered around hot processes, for example, furnace operations or welding. In this particle size range, it cannot be assumed that all particles inhaled will be deposited in the human airways. Experiments have shown that a minimum of deposition occurs at around 300 nm, with the majority of inhaled particles being subsequently exhaled.[102] Particles smaller than 300 nm are captured through diffusion onto surfaces. However, the capture surface is as likely to be in the head airways as in the lower lungs.

Certain impactors, such as the micro-orifice uniform deposit impactor[103] (MOUDI or Nano-MOUDI; MSP Corp., Shoreview, MN,) or the electrical low-pressure impactor[104] (ELPI; Dekati, Ltd., Tampere, Finland) can be used to measure size fractions in the sub-micrometer range. Stage deposits are typically analyzed chemically, although given the low significance of mass in this size range, the stage deposits may also be examined microscopically or by other techniques. Instruments that measure sub-micrometer particle numbers include the condensation nucleus (or particle) counter[105] (CNC; TSI, Inc, Shoreview, MN; Kanomax US, Andover, NJ; Met One Instruments, Inc., Grants Pass, OR) and the differential mobility analyzer[106] (DMA; TSI, Inc.). The CNC works by condensing liquid on fine particles until they are large enough to be determined optically; therefore it cannot provide information on the original sizes of the particles. The response varies with the nature of the aerosol and the condensing vapor; hygroscopic aerosols such as sodium chloride are detected better with water vapor, while non-hygroscopic aerosols such as silver and tungsten oxide particles are detected better with an alcohol vapor. A DMA, such as the Electrical Mobility Spectrometer VIE-08 (Hauke, GmbH, Gmunden, Germany), works by charging the aerosol and measuring the ability of particles to cross an electric field. Size distribution information can be obtained by changing the field intensity. More recent instruments combine the DMA with a CNC to produce the Scanning Mobility Particle Sizer (SMPS; TSI, Inc.). Other direct-reading instruments, such as time-of-flight laser optical particle counters have a range that extends partly into the sub-micrometre range. One of the problems encountered in using these techniques for ultra-fine aerosol monitoring is that the different instruments can give different interpretations of their measurements in the overlap of their ranges.[107] This is often a result of the methodology used to derive the aerosol size distributions from the actual measurements, a process known as inversion. This aspect must be controlled in combination instruments such as the wide-range particle spectrometer (MSP Corp.) which incorporates laser light scattering as well as a CNC and a DMA.

2.3.9   Calibration and Quality Control

Since the most common practice in air sampling is to calculate the result as a mass concentration of aerosol per unit volume of air, the volume of air flowing through the filter must be known. The size-selectivity of samplers varies with the aspiration velocity. If a specific size-selective fraction of the aerosol is intended as the sample, the flow-rate needs to be constant within a narrow range of variation around the specific flow-rate required for the desired collection characteristics. Accurate calibration of the air-flow through the sampling device is necessary. Placing the calibration device before the filter will give the most accurate result because in the connection between the filter and the pump the pressure is lower than ambient, and this may affect the read-out of the calibrator. Calibration devices should be traceable to primary standards. A primary standard in metrology has the specific meaning of national or international measurement reference standards. Traceability is demonstrated through a documented, usually annual, check of the measurement device using methods which are themselves traceable to the primary standards. Many of those who take air samples have not used the term "primary standard" correctly. For example, a burette in which there is a soap-bubble used with a stopwatch has often been referred to as a primary standard for flow-rate calibration. However, without an annual calibration of these two devices against standards which are further traceable to national or international standards, the calibration should not be referred to as "primary", no matter how precise. Modern electronic calibrators are becoming popular, and if annual recertification of calibration traceability is required, this is often accomplished by returning the devices to the manufacturer. Standard procedures for the calibration of air sampling pumps and their associated media are available from ASTM International (West Conshohocken, PA).[108]

Aerosol samplers may be machined from metal or formed from plastic. In either case, the tolerances in the manufacture should be sufficiently tight that the final product performs as anticipated. Poorly made products may not fit together properly or securely, and attempting to substitute parts from different manufacturers is usually not a good idea. A classic example is the fit of plastic cassettes onto metal cyclones. Plastic cassettes are manufactured by several different organizations, but metal cyclones are manufactured by different organizations to fit cassettes from just one of these manufacturers. Cassettes from other manufacturers may fit too loosely or the fit may be so tight that the plastic cracks. Even for a single manufacturer, the component parts of a product may not fit together properly. Leakage may occur in two ways. There may be leakage *in the sampler* so that the air passing through the filter is more than that passing through the entry orifice. This is unlikely to be noticed during calibration. Either the extra airflow is included in the calibration (*e.g.* if the sampler is placed in a jar), in which case the flow through the entry orifice may be less than required for optimum size-selection, or it will not be included in the calibration (*e.g.* if the calibrator is attached directly to the entry orifice) in which case the leaking air may bring additional particles to the filter. For the plastic closed-face cassette (CFC) sampler described below, a hand-held vacuum pump can be used to ensure a tight fit of pieces. However, this test will not determine whether there is leakage *around the filter*, which may be observed by a more pronounced ring of dust near the edge of the filter or a discoloration of the support pad. Even relatively clean ambient air contains ultra-fine particles, but these are effectively removed by air sampling filters. A tight seal around the filter will thus ensure the ultra-fine particle concentration in the outlet of the sampler will be much less than that entering the sampler. A test for leakage around the filter can be set up simply using a condensation nuclei counter (see Section 2.3.8) to determine whether a sufficient fraction of ultra-fine particles are removed from the ambient air.[109]

High-quality manufacture is especially important for size-selective samplers, since their selection characteristics will vary with various critical dimensions. The manufacturer should have a quality system in place to ensure those tolerances are maintained. It is possible for poor manufacturing quality to result in individual products of the same part number with very different characteristics.[110,111] For the personal coal-mine dust cyclone samplers sold in the US, federal law provides for the right of appropriate government agencies to "have their qualified personnel inspect each applicant's [manufacturer applying for a certification] control-test equipment procedures, and records and to interview the employees who conduct the control tests".[112] ISO 9000-series certifications require consistency in manufacture, but do not assure the resulting product is necessarily fit for use. For all aerosol sampling devices sold in the US, there are no certifications requiring the testing of manufacturing quality through performance, and no products known to the author are routinely tested by a third-party in a manner similar to detector tube certification, and so customers must either trust the vendor or arrange for their own tests. However, a pilot program where a single cyclone was circulated around multiple testing laboratories indicated that not all laboratories would necessarily produce substantially similar results.[113]

Wear can have an impact on the performance of re-usable samplers. The most obvious problem is in the deterioration of seals, especially o-rings, with time and use, but exposure to corrosive aerosols or cleaning products can also damage both metal and plastic parts. Finally, it is also necessary to consider the possibility of deliberate tampering (either by the employer or the employee).[114,115] For this reason, the MSHA coal mine dust sampling cassette[116] has been re-designed several times to incorporate tamper-proof features, including tamper-evident seals, one-way valves and diversion of the sample deposit to areas of the filter that are less easily accessible.[117,118]

2.4   Filters

The two main types of filter are glass or quartz fiber, and polymeric membranes including polyvinyl chloride (PVC), mixed cellulose ester polymers (MCE), polytetrafluoroethylene (PTFE), and polycarbonate (sometimes referred to as TMTP, although this is only a part of one manufacturer's part number). Filter porosity is determined from the ability to retain particles in liquid suspension and other mechanisms, including impaction, interception and electrostatic attraction, apply to filtration from air and are effective in removing small particles from the air stream through the filter. This results in the somewhat counter-intuitive recommendation to use filters with, for example, a 5 μm porosity to capture 1 μm aerosols.

Liu *et al.* reported tests performed on products available at the time showed many filters to be better than 95% efficient at capturing particles that are larger than 0.1 μm aerodynamic diameter, even when the pore-size is as large as 8 μm.[119] However, those results relate to products on the market 30 years ago; even if the same product is still available today, it may not be identical.

One method already noted as helping to achieve greater analytical sensitivity for chemicals with low target concentrations, such as beryllium, is to collect higher volumes of air. Increasing the flow-rate through a filter causes an increase in the pressure drop across it, which may eventually be beyond the capacity of the sampling pump, especially a personal pump. Increasing the filter porosity reduces the pressure drop. Hence it may be possible to pull $10\,L\,min^{-1}$ through an 8 μm MCE filter, even where it is not possible to pull a similar flow-rate through a 1 μm filter of the same type. The different filters vary in cost, ease of handling, and suitability for the preferred analysis.

2.4.1 Glass and Quartz Fiber Filters

Glass-fiber filters are the least expensive and have been in use longest. They are commonly used in gravimetric analyses in Europe, but not in the US. Their weight stability is not quite as good as that of PVC filters, but it is considered adequate in the countries where used.[120] They can also be used in methods requiring digestion of particulate as long as they are binder-free, and there are no analytical interferences. However, fibers may be released that might cause mechanical problems (*e.g.* by blocking aspiration tubes and nebulizers). Hydrofluoric acid can be used if necessary to completely dissolve the fiber.

Quartz fiber filters are more expensive than glass fiber filters. They are mostly reserved for special applications, for example in sampling diesel exhaust, where organic contamination must be first removed from the filter by baking at a high temperature.[121]

Glass and quartz fiber filters offer little resistance to air flow and thus they are the most common filters for high flow-rate applications. Whatman #41(Whatman plc, Maidstone, UK) is a typical glass-fiber filter.

2.4.2 PVC Filters

These are the filters most commonly used for gravimetric analysis in the US because of their exceptional weight stability, even though their price is significantly higher than glass fiber filters. It is possible to perform some additional chemical analyses on particles collected on PVC filters, although PVC is difficult to digest. Methods for the analysis of silica sometimes call for the PVC to be digested in the organic solvent tetrahydrofuran, or destroyed by plasma ashing.[122] Strong acid digestion generally leads to a very viscous matrix. Particles may be efficiently removed from PVC filters by ultrasonic agitation and then digested, but this procedure has not been extensively evaluated.[123] Filters that are a mixed polymer of PVC and polyvinyl acetate are also available.

2.4.3 MCE Filters

MCE filters are most commonly used for metals analysis as they dissolve easily in oxidizing acids such as nitric or perchloric.[124] The standard nominal pore-size is $0.8\,\mu m$, but this is not compatible with high flow-rate sampling. Larger pore-sizes are available, but the efficiency of collection of small particulates may be compromised if the pores become too large. A pore-size of $1.2\,\mu m$ is available and may allow higher flow-rates with low particle losses. The filters are hydrophilic making them a poor choice for gravimetric analysis,[125] but the attraction of being able to measure total particulate mass collected as well as perform chemical analysis causes them to be used this way in some European countries, *e.g.* Sweden.[126] The cost of MCE filters is between that of glass-fiber and PVC filters.

2.4.4 Polycarbonate Filters

These filters have the smoothest surface most suitable for microscopic analysis. They are used for optical microscopy, frequently for bioaerosols,[127] and transmission electron microscopy, *e.g.* for asbestos.[128]

2.4.5 PTFE Filters

These filters are the most expensive, but also the most inert, hydrophobic and free from interferences. They are often used for sensitive chemical analyses of reactive compounds, such as polycyclic aromatic hydrocarbons (PAHs).[129] Although they are prone to static problems, with careful handling they can be used to drive gravimetric analysis to a sensitivity of 1 or 2 µg per sample.[130]

2.4.6 Filter Support

Most filters require additional support against the vacuum from the pump. Although glass-fiber filters are relatively thick, they may also require support. Cellulose and porous plastic support pads are available, and some samplers have their own metal or plastic mesh supports.

2.4.7 Filter "Handedness"

Through their method of manufacture, polymeric filters tend to have one side shinier (less rough) than the other. Particles on the shiny side are more likely to be in the same plane (for microscopic or other surface analysis) and will be more easily removed for chemical analysis. Conversely, the rough side may have a larger overall capacity for collection, and collected particles may be less easily dislodged during transport and handling.

2.5 Samplers for Inhalable Sampling

Inhalable samplers can differ markedly in design, but all are intended to pull the ambient air through a filter or foam, thereby catching the aerosol on the filter or in the foam. A current issue is the difference between inhalability defined at the point of particle aspiration into the sampler and inhalability defined at the point of collection on the filter.[47,131] Some of the "inhalable samplers" described below are shown in Figure 2.1.

2.5.1 IOM Sampler

Only one commercially available sampler has been designed *a priori* to meet the inhalable convention.[83] This sampler was developed at the Institute of Occupational Medicine in the UK and thus is known as the IOM sampler (SKC, Ltd, Blandford Forum, UK). The 25 mm filter is housed in a capsule that fits

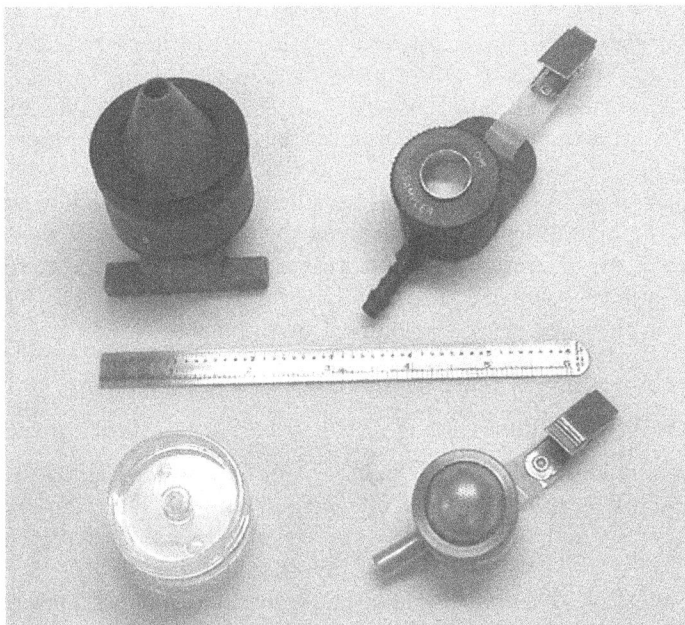

Figure 2.1 Inhalable samplers. Clockwise from top left: GSP; IOM; Button; CFC.

into a holder. The capsule has a 15 mm orifice that is held by the holder with the orifice forward from the body. It is operated at $2 \, L \, min^{-1}$. The capsule can hold either a glass fiber or membrane filter, with support from the base of the capsule.

Deposits on the internal walls of the capsule were noted in the development of the IOM sampler, and were considered sufficiently significant to require their inclusion as part of the sample.[132] Hence the capsule is weighed with the filter as a single piece in gravimetric analysis.[133] However, it is less clear what the procedure should be when the sample is to be analyzed other than by weighing. UK methods for metals, which specify the use of an inhalable sampler, and which reference the IOM sampler as one such, do not address the recovery of wall deposits for analysis.[134] Currently, neither do ISO standards,[135] but this is under review.

The capsule comes in two materials – a conductive plastic or stainless steel. The conductive plastic capsule has weight stability issues, which cannot easily be addressed by equilibrating the capsule at constant humidity, as this can take many days to achieve.[136,137] One recommendation is to carry several blank capsules (a minimum of three) to the field and correct samples for the mean change in weight of these blanks.[138] However, with just three blanks, the limit of quantitation is 0.5 mg. The stainless steel capsule is an alternative, but the weight of the capsule cannot be accommodated by many balances weighing to 1 µg, and it is considerably more expensive. The wide orifice also causes a dependence of collection efficiency on orientation to the wind, but this is obscured in tests which average wind directions.[139]

The front-facing wide orifice has been criticized for its ability to collect particles larger than the current limit of the inhalable convention by sedimentation and its ability to collect projectile particles of any size.[140–144]

The cost of the capsules means they must be re-used, and this may cause problems of cross-contamination. Cleaning the capsules is an additional cost in their use.

Nevertheless, the IOM sampler has been used extensively, both in laboratory and field studies, and its good performance within the limits of the inhalable convention in the laboratory studies has earned it a reputation as a "reference" inhalable sampler such that attempts have been made to relate IOM measurements in the field to the development of "inhalable" OELs.[87,145,146]

2.5.2   Button Sampler

The Button sampler (SKC, Inc. Eighty Four, PA) was designed to have an even deposit of sample across the filter, mainly for the direct observation of microorganisms, and for the collection efficiency to have a low dependence on wind speed and orientation.[147] It was not designed to be an inhalable sampler, but limited laboratory studies suggest performance may be within the criteria laid down in EN 13205:2001, although this has not been explicitly tested.[28,148]

The Button sampler has some practical advantages over the IOM sampler. Without internal walls, there are no wall deposits to be accounted for, and with a screen of small holes acting as the entry orifice, the entry of very large or projectile particles is minimized.[149] The chief drawback to the Button sampler is the very high flow-rate required to meet the inhalable convention (4 L min^{-1}) coupled with the use of a small (25 mm) filter, which results in a very high pressure drop across a membrane filter, and especially in atmospheres with any moderate dust concentrations. The use of glass fiber filters therefore is recommended by the manufacturer to avoid the high pressure-drop and consequent reduction in pump capability. The Button sampler is relatively expensive and thus not disposable.

2.5.3   GSP Sampler

The Gesamtsstaubprobenhame, or GSP sampler (Gesellschaft für Schadstoffmesstechnik, GmbH, Neuss, Germany) has a conical inlet flaring from a 9 mm orifice to a 37 mm filter in an inert plastic housing. This conical inlet has caused it to be colloquially referred to as a "cone" sampler. Other conical samplers exist, and assuming the dimensions are exactly equivalent, they may have similar aspiration efficiencies.

The GSP is made of metal, but at least one alternative is constructed of conductive plastic. The GSP sampler was developed to have a specific inlet velocity, and a number of inlets are available to allow different flow-rates to achieve this velocity. The standard inlet cone design used commonly in

Germany[150] requires a relatively high flow-rate of $3.5 \, L \, min^{-1}$. However, since the filter is 37 mm diameter, the pressure drop even for membrane filters is manageable for most good quality personal sampling pumps. While not designed explicitly for the inhalable convention, the performance of the GSP sampler has been found to match it reasonably well.[85] Wall deposits are not included in protocols using the GSP sampler, although they do occur.

Although it is not well-known in the US and there is no US distributor, the German manufacturer can supply directly to the US market. As with many other samplers, the cost makes re-use a practical necessity.

2.5.4 CFC Sampler

A plastic casette made from polystyrene or poly(styrene-acrylonitrile) is the most commonly encountered sampler in the US, and also in several other countries. The advantage of this device is its low cost, which allows single use, and, therefore, freedom from cleaning, potential cross-contamination, and wear. It is available in different sizes to contain either 25 mm or 37 mm filters, and can be used open-faced or closed-faced; in the latter case, the entry orifice is 4 mm diameter. In open-faced mode the deposit is generally even but also very open to tampering or accidental disruption. It is more difficult to tamper with or disrupt the filter in the closed-faced mode, but the deposit is then unevenly distributed across the filter, often being concentrated in a small mound directly under the orifice. Even in the closed-face mode it is possible to tamper with filter deposits, and so a "tamper-proof" version of the sampler exists, as described above. The most common version in use in the US is the 37 mm closed-face cassette (CFC). It is manufactured by many companies today, but is also still available from the original manufacturer (Millipore, Corp., Billerica, MA).

2.5.5 Evaluating Internal Wall Deposits

The entry inlet and the flow-rate of the IOM sampler were optimized during development for a specific probability of size-fraction collection. The "sample" of interest therefore was regarded as all particles which passed through the entry orifice. Many airborne particles possess a static charge which can induce an opposite charge in the sampler walls leading to an electrostatic attraction. Large particles may be projected through the orifice and impact the walls of the capsule containing the filter or fall through the orifice and onto the walls by gravitational settling.[151] They may also bounce off the filter and back onto the walls.[152] During testing of the IOM sampler, it was noted that the proportion of particles passing the orifice being deposited on the walls was a significant part of the total sample and therefore needed to be included, and so the change in weight of the capsule as a whole was considered the collected mass.[132] Other studies since that time have confirmed the deposition of particles on the walls of the IOM sampler. Table 2.1 summarizes data from field studies for metals where wall deposits have been accounted for separately. Studies where

Table 2.1   Median and maximum wall deposits in the Institute of Occupational Medicine filter cassette (SKC, Inc.) encountered in various aerosol measurement studies.[a]

Environment	n	Agent	Median Wall Deposit (%)	Maximum Wall Deposit (%)
Cu smelter[131]	17	Cu	16	38
Pb ore mill[155]	8	Pb	19	30
Battery production[153]	11	Pb	8	33
Welding[153]	18	Al	3	13
Cast iron foundry[154]	18	Fe	8	69
Grey iron foundry[154]	18	Fe	5	16
Bronze foundry[156]	6	Cu, Pb, Sn, Zn	0, 0, 0, 3	10, 3, 23, 6

[a]Wall deposit as a percentage of total (filter + walls) sampler catch.

metal-containing particles were rinsed from the walls[153,154] appear to show less deposit than those where metal-containing particles were wiped from the walls.[131,155] Thus rinsing alone may not be effective at removing all the wall deposits, and further evidence for this comes from laboratory[82,132,142] and field investigations[146] where the samples were weighed and where wall deposits were typically around 25-30% of the total aspiration of the sampler.

The same issue has arisen for the CFC. In the US, OSHA considers the sample to be all the particles that pass the orifice of the filter-holder. For total mass determinations (OSHA PV 2121),[157] OSHA uses a filter held in a capsule, (Woodchek™, MSA, Inc., Cranberry, PA) and the change in weight of the entire assembly includes any wall deposits in the sample. However, for analyses where the filter is digested for subsequent analyses, the internal capsule is not used.

Some early OSHA analytical methods for specific chemicals required the inclusion of wall deposits when visible. An increasing awareness that significant mass of sample may be on the internal surface of the cassette but not visible to the eye has caused OSHA to decide to analyze all samples for wall deposits. This has been made clear in new methods and revisions of existing methods. A summary of wall deposit evaluations for the CFC sampler is presented in Table 2.2. Where it has been determined that particles on the internal surfaces of the sampler should be included, the question arises as how they should be recovered and analyzed. Rinsing has been used, but this has been shown by OSHA to be inadequate for quantitative removal. Physical pressure is required to dislodge the particulate. Therefore, wiping with wetted fabric appears to offer the best alternative.[160] The fabric piece should be as small as possible so as not to interfere with the analysis, and the wipe should be compatible with the analysis. OSHA uses polyvinyl alcohol "Ghost" wipes routinely, although not in the case of hexavalent chromium, which can be easily reduced (PVC filters wetted with buffer are used instead).[159] Cellulose wipes may have lower backgrounds and might digest more easily. The wipe can be added to the filter for analysis.

Table 2.2 Median and maximum wall deposits in the 37 mm closed-face
plastic filter cassette encountered in various aerosol measurement
studies.[a]

Environment	n	Agent	Median Wall Deposit (%)	Maximum Wall Deposit (%)
Cu smelter[131]	18	Cu	21	55
Pb ore mill[154]	9	Pb	19	35
Solder manufacture[158]	30	Pb	29	74
Battery production[154]	16	Pb	28	66
Welding[159]	10	Cr(vi)	5	55
Plating[159]	12	Cr(vi)	12	17
Paint spray[159]	29	Cr(vi)	7	12
Zn foundry[154]	9	Zn	53	62
Zn plating[154]	18	Zn	27	91
Cast iron foundry[154]	18	Fe	22	46
Grey iron foundry[154]	18	Fe	24	77
Bronze foundry[156]	6	Cu, Pb, Sn, Zn	19, 13, 0, 15	45, 17, 0, 21
Cuproberyllium[154]	4	Cu, **Be**	31, **12**	40, **39**

[a]Wall deposit as a percentage of total (filter + walls) sampler catch.

If thought necessary, a second wipe could be analyzed separately to confirm removal of all particulate by the first. A requirement for wiping does add a further degree of complexity to the method evaluation. For example, in the case of beryllium, both soluble salts and insoluble materials may need to be tested for recovery efficiency. The OSHA study of recovery of hexavalent chromium[159] provides a good template for such a study. *In situ* digestion is another alternative, which is used by the INRS (France).[161] The digestion procedure calls for glass-fiber filters to be dissolved in an acid solution containing hydrofluoric acid. The acid solution is pipetted into the CFC, so that more than just the integrity of the air sampling process may be dependent on the tightness of fit of the cassette pieces. No large scale studies of *in situ* digestion have been published with the acid compositions commonly used in the U.S.A. Poor recovery might be aided by invoking ultrasonic or microwave energy. Certainly, for laboratories engaged in multiple routine samples there may be some considerable cost-savings to *in situ* digestion through not having to open each cassette, remove the filter and wipe the interior. The addition of acid through the entry orifice of the CFC should be amenable to automation. The pharmaceutical industry has developed a PVC capsule with PVC filter for *in situ* extraction with less aggressive solvents.[162] This same capsule with an MCE filter might be usable with acid digestion, where the filter would be dissolved but the cap would be left floating free. Alternatively a MCE capsule might be developed.

2.5.6   The CFC and the Inhalable Convention

The concept of the existence of large particles and their potential to be inhaled was not considered in the early evaluation of the CFC, for several reasons. First, large particles were considered to settle out of the aerosol too quickly to be important. Second, the experimental techniques to generate and calibrate large particle aerosols in the laboratory had only been developed recently. Third, when the metric is particle number, the presence of a few heavyweight particles is irrelevant. Thus early evaluations of the CFC used particles up to around 10 µm AED.[71] When later it was tested against the inhalable convention, the CFC was tested as it had always been used, with analysis only of the filter deposit.[85] The sampling efficiency was not considered a match to the inhalable convention in these experiments. Later, it was realized that internal wall deposits in the CFC are a significant part of the aspiration. Since the CFC has been shown to sample smaller particles rather efficiently, it seems likely that these wall deposits are larger particles. Wall deposition in the IOM sampler was shown to increase with increased particle size.[85,132,142] However, there has not been any investigation of particle size difference between deposits on the walls and on the filter of the CFC.

When the mass of material on the walls of the CFC is added to that in the filter deposit, for samples taken side-by-side with the IOM sampler in various metals industries, it has been shown that differences between the CFC and the IOM "reference" inhalable sampler become much smaller.[163] The CFC with wall deposits included has not been tested in the laboratory against the inhalable convention, and this test should be carried out since the CFC has important advantages over other samplers in terms of cost and disposability, and ease of calibration and deployment that must also be included when considering practical field use. The GSP sampler is likely to have wall deposits similar in degree to those of the IOM and CFC, but this has not been fully evaluated. The Button sampler, however, has no "walls" as such. It should be noted that no analytical procedure that interrogates the filter only, such as X-ray fluorescence for lead, can include wall deposits.

2.5.7   CIP-10 Sampler

The CIP-10 sampler operates differently from other samplers. The sampler rotates at high speed pulling air through porous polyurethane foam by a centrifugal effect.[164] The foam acts like a filter and retains particles from the air stream. The absence of a significant pressure drop across the foam allows the sampler to work efficiently with small batteries. However, calibration becomes more complex and requires a laboratory bench set-up. Once calibrated to a particular rate of air movement, however, the speed of rotation of the sampler can be measured and used as a field check of air flow. The foam is held in a cup into which some larger particles can fall. Therefore the foam and cup are weighed together to determine total mass. Particles in the foam can be extracted for alternative analyses by sonication of the foam in liquid. Particles in the cup must also be added. The CIP-10 sampler can be used with heads designed for specific size collection (inhalable, thoracic and respirable). Nominal flow-rates are 7–10 L min^{-1}, depending on the head selected. The CIP-10 is available directly from the manufacturer (Arelco, ARC, Fontenay-sous-Bois, France).

2.5.8 An Inhalable Convention for Slowly Moving Air

Researchers developing size-selective sampling conventions over-estimated the wind speeds typical of indoor workplaces. The inhalable convention, although it was based on test data to 0.5 m s^{-1}, is actually stipulated as relevant to wind speeds above 1 m s^{-1}. In reality, wind speeds measured in a variety of workplaces have often found to be less than 0.3 m s^{-1}.[79] This situation has been referred to as still air or calm air, but it should more correctly be called "slowly moving air". Air movement factors into the equations of aerosol behavior and it is no surprise that aerosol behavior may change as the wind speed approaches zero. A proposal has been made for the application of different inhalability criteria for slowly moving air based on manikin studies,[80] in which the IOM and GSP appeared to compare well with the newly proposed criteria.[81] Two other studies include statements relevant to this situation. The report of the major European study of sampler performance concluded that analysis of the filter deposit only of the IOM may better meet the current inhalability criterion for their lowest wind-speed test condition (0.5 m sec^{-1}),[85] while in an investigation of the IOM sampler for large particles in very low wind speeds (<0.3 m sec^{-1}), the filter-only catch better matched the proposed convention for inhalability in slowly moving air.[82] Further investigations are evidently needed.

2.5.9 Very Large Particles

The ISO inhalable convention does not pass through zero aspiration efficiency at any size, although it is obvious that there is a certain upper limit of particle size that cannot be inhaled effectively. As particles increase in size, their stopping distance increases. A short stopping distance means small particles tend to be carried by air currents, while longer stopping distances tend to cause larger (or faster moving) particles to impact on surfaces. The gravitational settling rate also increases for larger particles. The combination of these effects results in an upper limit to the size of particles likely to be present in workplace air,[165] although projectile particles can cross relatively large distances to the worker.[140] Unfortunately, it is difficult to perform experiments on both human inhalability and sampler performance with very large particles. The ISO convention stops somewhat arbitrarily at 50% efficiency at 100 μm AED. More recently, experiments have been carried out concerning the inhalability of larger particles,[166–169] and also on the performance of samplers in collecting large particles.[142,144] The conclusion to be drawn from these studies is that, while inhalability of particles falls to near zero somewhere between 80 and 150 μm AED, the IOM sampler collects these very large particles with efficiency greater than 50%, increasing with size. Given the cubic relationship between mass and diameter, it takes very few such particles to dominate the total sample collected, while their actual contribution to dose may be rather small. Samplers with smaller inlets, including the CFC and Button samplers collect fewer very large particles, and their use should be considered in workplaces where very large particles are common such as woodworking[170] or high-speed metal machining.

2.6   Samplers for Respirable Sampling

Most of the respirable samplers available feature a cyclone size-selector fol-lowed by a filter collector. Coarser dust falls out in the body of the cyclone and the filter catch is assumed to be the sample. At least a half-dozen different designs of cyclone are commercially available. Many of these connect to a filter housed in the CFC as described above. Wall losses can also occur in these types of samplers and may need to be accounted for.[171] The major differences between the different types of cyclones are inlet design, body-shape and flow-rate. The inlet design also influences the manner of calibration.

All cyclones have size-selection efficiencies that differ from the particular respirable convention (and also from each other). In general, the cyclone separation curves are steeper than the convention curves, leading to an over-sampling of smaller particles and an under-sampling of larger particles. Since workplace aerosols generally encompass a wide range of size distributions, this may result in a partial cancellation of effects.

Once the performance curve of the cyclone is known, the sampling efficiency can be compared to an ideal "perfect" cyclone for any reasonable aerosol distribution that is log-normal. Then an overall bias between the cyclone and ideal can be calculated for any geometric mean (GM) and geometric standard deviation (GSD) of a log-normally size distributed aerosol, and this can be plotted and contours of equal bias drawn on a "bias map".[172] A guide to sampler performance is to determine the likely bias for the aerosol under study, although, in practice, sampler development is guided by the more general goal of minimizing the bias over the whole space of reasonable values of GM and GSD. These principles have been employed recently in testing a range of samplers.[173] Changing the flow-rate through the cyclone moves the slope of the separation curve according to a fairly simple model that allows prediction of the performance of the cyclone at different flow-rates. Performance can thus be extrapolated between different flow-rates to determine the optimum flow-rate (being defined as that which minimizes the overall bias of the GM/GSD space and not where the 50% efficiency of the cyclone is closest to the 50% penetration efficiency, or D_{50}, of the convention). Many cyclones therefore have the capability to determine concentrations based on more than one size-selective convention with simply a change in the flow-rate. Some of the "respirable samplers" described below are shown in Figure 2.2.

2.6.1   Comments on Cyclone Design

Entry orifices of cyclones may be holes or otherwise shaped cuts in the exterior of the body leading directly into the vortex chamber, and these are known as

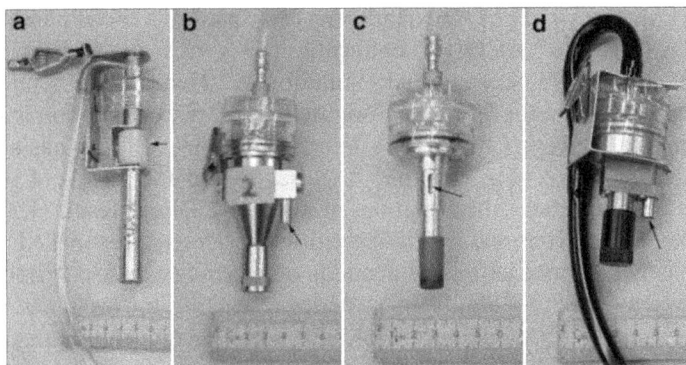

Figure 2.2 Respirable samplers. Left to right: DO cyclone; GK 2.69 cyclone; aluminium cyclone; Higgens-Dewell cyclone. Photo courtesy Paul Baron,
NIOSH.

"open" inlet cyclones. Alternatively, the entry inlets may be tubes typically
opening to the air downwards and passing through a right-angle before
entering the vortex chamber, where the cyclone is often referred to as having a
"blind" inlet. The latter is less susceptible to the influence of wind strength and
direction. Attempts to reduce wind effects on open-inlet cyclones include
increasing the number of inlets and their spacing around the cyclone.

Cyclone bodies are generally either long and thin or short and squat. Longer
cyclone bodies allow more internal air turns, but may also result in instability in
the number of turns. Cyclones may be of conductive plastic or metal to avoid
particles sticking to the walls by electrostatic attraction. It will be noted in the
descriptions below that there are conflicting conclusions with regard to the most
appropriate flow-rate for cyclones to meet specific size-selective conventions.

Calibrations may differ because of the particles chosen. For example poly-
styrene-latex (PSL) spheres are round and bouncy, while potassium sodium
tartrate crystals have a distinctive shape and less rebound on impact. Oleic acid is
a very different, liquid aerosol. The particles used may have a very narrow size
distribution ("monodisperse") or a very wide size distribution ("polydisperse").
Generation systems, exposure chambers (*i.e.* wind tunnel *vs.* calm air), and
detection systems may also differ. It is perhaps not surprising that calibrations
may differ given the wide choices available and lack of standardization. An
attempt to research this issue further by passing a single cyclone between
laboratories for calibration regrettably was not funded beyond the pilot stage.[174]

2.6.2 The Dorr-Oliver (DO) or "Nylon" Cyclone

This cyclone is manufactured by several organizations (*e.g.* Sensidyne, Inc.,
Clearwater, FL; MSA, Inc., Pittsburgh, PA). It is made of non-conductive
nylon, and has a single slit inlet. The filter is held in a CFC attached through a
connection at the top of the cyclone. The DO cyclone is difficult to calibrate,

requiring a special jar. The DO cyclone was calibrated to a specific size-selective convention (the "old" ACGIH® convention, $D_{50} = 3.5\,\mu m$) and this convention is specified in US occupational regulations.[175] Hence, it is the most commonly used cyclone in the US. The flow-rate to meet this convention is specified at $1.7\,L\,min^{-1}$. However, based on a later calibration, NIOSH has suggested that the performance at $1.7\,L\,min^{-1}$ is closer to the ISO ($D_{50} = 4\,\mu m$) convention.[172] Yet another calibration led to an even different result ($1.3\,L\,min^{-1}$ to meet the ISO convention).[176] This situation has yet to be reconciled. The DO cyclone has been criticized on the grounds of sensitivity to wind strength and direction,[177] and the possibility of particles impacting on the walls opposite the entry leading to a deposit build-up with consequent change to the size-separation.[176,178,179]

2.6.3   The GS-3 Cyclone

The GS-3 cyclone (SKC, Inc., Eighty Four, PA) is named for Gautam and Sreenath, researchers at West Virginia University, US. It was designed to avoid some of the criticisms noted for the DO cyclone.[177] To reduce sensitivity to wind, there are three slit inlets at 0, 90 and 180° (a fourth at 270° is unnecessary as it would face the wearer's body) and the inlets are more tangential to the body, which prevents deposit build-up by impaction on the opposite wall from the inlet. In addition, the material is static-dissipative plastic, thus preventing further internal deposition from electrostatic interactions. It operates at a flow-rate of $2.75\,L\,min^{-1}$ to meet the ISO ($D_{50} = 4\,\mu m$) convention, and the calibration data have been extrapolated to suggest a flow-rate of $3.5\,L\,min^{-1}$ to meet the "old" ACGIH® convention ($D_{50} = 3.5\,\mu m$).

This is a long-bodied cyclone. As with the DO cyclone, the cyclone is connected to a filter holder, which, for consistency, should be static-dissipative like the GS3 itself. The GS3 also requires a jar for calibration. A single slit version (GS-1) is considered by the manufacturer to be "equivalent" to the DO cyclone. However the recommended flow-rate ($3\,L\,min^{-1}$) to achieve the "old" ACGIH® D_{50} of $3.5\,\mu m$ is very different from that of the DO cyclone. This suggests there might be a difference in performance at other particle sizes sufficient not to convince regulatory authorities to accept the cyclone as meeting this convention.

2.6.4   IOSH Cyclone

This is another version of the DO cyclone in conductive plastic developed through the Institute of Occupational Health (IOSH) in Taipei, Taiwan (Republic of China) to be a close match to the ISO respirable convention over the entire range of the curve, and the closeness of the match has been demonstrated in the laboratory and the field.[179,180] Like the DO cyclone it must be calibrated in a jar, and it is operated at $1.7\,L\,min^{-1}$. It is available in the US from New Star Environmental, Inc. (Roswell, GA).

2.6.5 Aluminium Cyclone

The SKC, Inc. (Eighty Four, PA) aluminium cyclone was based on a Scandinavian design to meet the Johannesburg Convention ($D_{50} = 5\,\mu m$), and the flow-rate for this purpose was $1.9\,L\,min^{-1}$.[181] It was later re-characterized for the ISO convention ($D_{50} = 4\,\mu m$), with a flow-rate of $2.5\,L\,min^{-1}$.[182] A more recent calibration suggested $2.2\,L\,min^{-1}$.[176] It has a long slit inlet, but yet it is relatively easy to calibrate. The filter is held in a plastic cassette in open-face mode. Versions of the cyclone fitting both 25 mm and 37 mm filter holders are available.

2.6.6 Higgens-Dewell Cyclone

In the US, BGI, Inc. (Waltham, MA) is a source for a cyclone based on a design from the UK.[183] It has a downward opening tube inlet that is easy to connect to a calibrator. It is available in both nickel-plated aluminium or conductive plastic. It was originally designed for the BMRC convention ($D_{50} = 5\,\mu m$ AED) with a flow-rate of $1.9\,L\,min^{-1}$, but has also been shown to meet the ISO convention ($D_{50} = 4\,\mu m$ AED) at a flow-rate of $2.2\,L\,min^{-1}$.[172]

2.6.7 GK2.69 Cyclone

Also available from BGI, Inc. (Waltham, MA) is a wide-bodied cyclone that meets the ISO ($D_{50} = 4\,\mu m$) convention at a relatively high flow-rate of $4.2\,L\,min^{-1}$.[184] It also has a downward pointing tube inlet that is easy to connect to a calibrator. It is available in anodized aluminium, or stainless steel for corrosive atmospheres. This cyclone will also meet the ISO thoracic convention at a flow-rate of $1.6\,L\,min^{-1}$.

2.6.8 FSP-10 Cyclone

The highest flow-rate personal cyclone currently available is the FSP-10 (Gesellschaft für Schadstoffmesstechnik, GmbH, Neuss, Germany), which operates using a personal sampling pump at $10\,L\,min^{-1}$.[185] In order to ensure continuous operation without fault due to pressure drop across the filter, the manufacturer recommends a large nominal pore-size filter (8 μm).

2.7 Sampling for Different Fractions

Several newer devices claim to be able to sample more than one of the ISO sampling conventions simultaneously.

Size-selective polyurethane foam inserted into the throat of an IOM sampler is meant to catch the larger particle sizes in the inhalable fraction, allowing the respirable particles to penetrate to the filter.[186] It was necessary to redesign the capsule with a longer throat to accommodate these foams. The effect of this change on the aspiration efficiency may be minimal.[187] A further change has been made to include an internal lip to prevent foam from being pushed down onto the filter. This may not affect the aspiration efficiency, but it may affect the distribution of particles between capsule walls and filter (all data in Table 2.1 are from capsules without the internal lip). Weighing the entire sampling capsule, including foam and filter, gives the inhalable fraction of the aerosol, while weighing the filter catch alone gives the respirable fraction. Changes in particle size penetration with sample loading were noted when similar samplers were used to collect welding fume,[188] and were also noted in more recent laboratory and field tests,[28] but this occurs when loadings are in the range of several milligrams of dust, and sampling in areas where beryllium is an issue may collect rather less dust.

The Respicon sampler (TSI, Inc., Shoreview, MN) consists of three separation stages (virtual impactors) under an inhalable inlet.[189] The final filter (Stage 3) as usual provides the respirable fraction. Adding the mass on the next filter (Stage 2) gives the thoracic fraction. Adding the mass on the first filter (Stage 1) to that on the other two gives the inhalable fraction. It has been tested for wood dust and in nickel production.[152,190,191] It operates at $3.1 \, L \, min^{-1}$. It is relatively expensive.

Marple-style personal cascade impactors, such as the Series 290 (Tisch Environmental, Village of Cleves, OH) are compact and lightweight, and are available with eight, six, or four impactor stages all followed by a built-in filter holder.[192,193] The impactor stages are typically special filters, oiled to prevent particle bounce (although bounce is still possible), and therefore loss to subsequent stages. At the design flow-rate of $2 \, L \, min^{-1}$, the Series 290 has stage cut-points that range from 21 to $0.5 \, \mu m$. An inversion routine provides the aerosol size distribution from analysis of the mass collected on each stage. From this distribution, the mass fraction of respirable (any convention) or thoracic aerosol can be calculated in theory, although obtaining accurate results has been more difficult in practice. All particles on the first stage must be assigned to a "bin" between the cut-off of the first stage and the aspiration efficiency of the sampler, which might not be well known. This makes accurate assessment of the inhalable fraction impossible.

The Personal Inhalable Dust Spectrometer (PIDS) is a cascade impactor which has an inlet intended to meet the inhalable convention, and an upper stage consisting of porous polyurethane foam with a cut-point around $30 \, \mu m$, that provides some discrimination in the extra-thoracic fraction.[194] However, this unit is not currently commercially available. While impactor data can be used to model the different size-selective conventions, with accuracy increasing with the number of stages, this does involve a trade-off for analytical sensitivity by additionally dividing the sample.

2.8   Sampling in Beryllium Facilities

A wide range of approaches has been used to collect the many air samples that have been analyzed for beryllium. Samples have been taken to determine compliance with limit values, but often have been taken for research purposes, for example in determining relative risks of job categories, tasks, or areas. A large part of the risk assessment process has involved discussions around the relative risks of different particle sizes, and this has also prompted the use of samplers to characterize different size fractions. A very good review is provided by Kolanz et al.,[195] although their summary position that air sampling was not a good predictor of risk for CBD will not be debated here. They do point out that many studies are not directly comparable because they used a variety of exposure assessment methods that are not necessarily representative of individual worker exposures. Future rationalization and consensus of procedures would be a useful goal.

Kolanz, et al. divided historical methods into "fixed airhead", "high volume" and "low-flow personal lapel".[195] The fixed airhead samples (also known as "continuous" samples by the US Department of Energy[196]) used 25 mm diameter Whatman 41 filter and a vacuum pump flow-rate of $10–20 \, \text{L min}^{-1}$ as area samples over long periods of time to characterize the source and variability of emissions, and the effectiveness of controls. The high volume samples were very high flow-rate, very short-term samples taken close to workers to evaluate task-based and peak exposures. High volume samples were taken at flow-rates of $200–400 \, \text{L min}^{-1}$ through a 10.5 cm diameter Whatman 41 filter with a vacuum pump as dictated by the needs for analytical sensitivity of the methodologies available in the 1950s. The results along with time and motion studies were used to calculate "daily working averages" (DWAs) for the workers, as well as to evaluate the need for and effectiveness of controls. The low-flow personal lapel samples were typically full shift CFC samples, which only became common in the 1970s after the development of the CFC and personal sampling pump. Studies comparing personal lapel sampling and area samples produced no good correlation, and an understanding that area samples underestimate personal exposure – lessons that have also been learned in many other industries. A study of the DWA procedure *versus* personal lapel sampling also showed higher results for the personal samples and poor correlation between the two exposure assessment methodologies.[197] Some efforts were put into investigating the causes of the differences observed between sampling procedures. Sampling on the lapel was not observed to give different results from sampling at the nose or forehead for well-mixed aerosols,[39] but an interference from particles re-suspended from clothing was noted.[42]

Particles larger than 10 μm AED tend to be among the largest contributors to dose measured in mass. Throughout one specific beryllium production plant under study, extrathoracic material made up 76–80% of the mass collected by a multistage impactor.[50] If it is assumed that the impactor inlet aspiration efficiency for large particles is less than that required to meet the inhalable convention, then the extrathoracic fraction of a perfect inhalable sample in this environment may easily be 90–95% of the sample mass. The respirable mass portion of such an inhalable sample would probably be less than 10%. Martyny et al.[51] examined beryllium machining operations and found that, in personal impactor samples, a median of 28% of the mass was in particles larger than

10 μm AED. Although the personal cascade impactors used side-by-side with CFCs gave consistently higher results, the analysis was only of the filter-catch of the CFC and did not include the wall deposits. Even the cascade impactor data should be taken as a low estimate of the extrathoracic mass. Kelleher *et al.*[198] also examined the exposure of machinists, and found that the larger particles (in this case, those >6 μm AED) dominated the samples in many cases (60–70%). Therefore, if the inhaled mass dose is considered to be the most important metric, it is logical to propose meeting the inhalable convention as the target for sampling. If samplers that have been tested to meet the inhalable convention are not practical in this application, then at least the interior wall deposits of the CFC can be included to get a closer approximation to the inhalable fraction than analysis of the filter alone.

At an ACGIH® Symposium on Advances in Air Sampling in 1987, Raabe presented a tripartite protocol for beryllium sampling, which included inhalable sampling for soluble beryllium salts such as beryllium fluoride and sulfate, thoracic sampling for the poorly soluble hydroxide and low-fired oxide, and respirable sampling for the relatively insoluble metal alloys, silicates and high-fired oxide.[199] Since large particles are expectorated and ingested, and only 0.2% of ingested beryllium was absorbed by the body, it was assumed that the inhalable convention would be important for highly soluble beryllium compounds, but not for the others. However, in an analogous situation, it has been suspected that the adsorption of even a small percentage of sparingly soluble large lead particles expectorated and ingested can outweigh the contribution to body burden of fine particles deposited in the alveolar region with much greater solubility.[200] The ACGIH® does not differentiate between different chemical species in proposing the inhalable convention for beryllium. The rationale for inhalable sampling is that all routes of exposure can contribute to sensitization, and thus all airborne particles are suspect.

There is also evidence that fine particles of beryllium have an effect in the alveolar region. The granulomas associated with CBD, which are found in the alveoli, appear to be associated with the deposition of particles containing beryllium on the alveolar walls. Thus correlations have been found between respirable mass and CBD, but also between respirable mass and sensitization.[86] If the mass of the respirable fraction is considered a better target for disease risk, then it would be no surprise that CFC sample results do not correlate well with risk due to the significant and variable impact of large particles on the mass collected.[86] In such case, it would be appropriate to select a respirable convention and a sampler, such as a cyclone, to meet this convention. While there is some suggestion that ultra-fine particle numbers might correlate better than any mass measurement with disease risk, further research is probably necessary to support this hypothesis.[201] The selection of an appropriate exposure metric for inhalation exposure to beryllium requires an improved understanding of factors such as chemical form, surface area and particle size that may affect the bioavailability of beryllium following exposure.[202] Until this is achieved, the inhalable convention might indeed represent the most conservative airborne fraction to prevent sensitization, even though the respirable fraction may be the most biologically relevant fraction to prevent progression to disease.

2.9 Sampling Emissions Sources for Beryllium

Sampling inside the workplace supports hygiene decisions within the workplace. Environmental sampling is used to demonstrate compliance with environmental standards or the need for action in the case of non-compliance. The most common pathway for the release of beryllium into the environment is through stack emissions, otherwise known as "stationary sources", controlled in the US through EPA regulations. While it is necessary to monitor emissions from facilities that produce and use beryllium, its compounds and alloys, it also necessary to monitor emissions from less obvious sources, for example, coal-fired power plants and aluminium refineries (beryllium is a natural contaminant of coal and bauxite). The current emissions limit is 10 g over a 24-hour period or 0.01 $\mu g/m^3$ averaged over a 30 day period (demonstrated for at least three years).[203]

The method used to sample stack emissions involves a heated isokinetic sampling probe, a heated filter in a borosilicate glass holder, and a series of impingers containing de-ionized water in an ice bath. The impingers are designed to collect any ultrafine particles passing the filter, and to condense any vapor-phase material. This is the standard "Method 5" sampling set-up.[204] Samples are usually taken to cover the maximum emission period within 24 hours, typically for a minimum of two hours at a flow-rate of 28 L min^{-1}. The two methods specific to beryllium are Methods 104 and 103. Method 104 references, and is similar to, Method 5, but it requires borosilicate glass probes rather than metal, and has procedural details specific to the analysis of beryllium.[205] Method 103 is a screening method also specific to beryllium, where the filter is not heated and the impinger train is not included.[206] The flow-rate is 14 L min^{-1}.

2.10 Analytical Considerations for Selecting a Sampling Method

Fully evaluated and documented sampling and analytical procedures are the gold standard. Evaluation includes a determination of the uncertainty around the final result. The uncertainty of a measurement is a function of variations in both the sampling and analytical steps. The uncertainty of the method is usually derived from laboratory tests[207] where, under carefully controlled conditions, an expanded uncertainty of 25–30% is usually allowed.[207,208] Uncertainty can also be determined from field evaluations against reference sampling and analytical methods, where an expanded uncertainty of 35% allows for environmental background variation.[209] In the sampling step, the only variation separately considered is that of the sampling pump. A flow-rate variation of 5% is typically used in method evaluation. Where size-selection is considered important, the difference between the collection efficiency of the sampler and that of an ideal sampler whose size-selective characteristics exactly match a reference convention can be calculated as an average bias for log-normal aerosol distributions over a range of geometric means and standard deviations within a realistic range, which can then be entered into the uncertainty calculation. If the true aerosol size distribution is known, a more precise estimation of the bias is possible.

Once the sample has been collected, and if it must be transported to a laboratory, it may be possible to assess handling and transport contamination or losses through examination of field blank samples. Field blank samples consist of sampling media taken to the field and assembled, and then immediately disassembled and sealed and transported to the laboratory. Field blanks should be included at a rate of 5% or 10% of samples, but at least one per day of sampling. Field blanks are generally reported separately, but may also be considered as a correctable bias, assuming they fall within specific tolerances.

In the laboratory, it is important to determine the analytical background of the media and any reagents used. This background is typically considered a correctable bias of low variance, so that the results are automatically corrected before presentation to the client. However, background may vary strongly between different manufacturers or between different lots from the same manufacture. Where the laboratory supplies the media for field sampling, they typically possess sufficient to be able to perform a media blank analysis. Where the field personnel have acquired their own media, it is important that they acquire sufficient to provide additional unused materials to the laboratory. It is also possible that the media may interfere with the analysis (for example, through interfering chemicals or fluctuations in media composition). Therefore, it is important for the laboratory to carry out a recovery study for the chemical of interest spiked onto the media. A large number of recovery studies may be necessary where recovery varies with lot, which encourages the production or purchase of bulk lots. Most recovery studies have involved only the filter deposit. Where aerosol deposits on the interior surfaces of the sampler are considered part of the sample a further recovery study may be necessary.

Recovery studies involve dosing the media with a known amount of analyte and processing the spiked media as if it were a field sample. The analytical result compared to the reference dose is the recovery. Recovery is usually less than 100%, but in certain conditions it can be more (*e.g.* where part of the solvent evaporates or is absorbed by the sample media). The study usually involves spiking multiple replicates at several different loadings encompassing that anticipated from sampling at a target concentration, usually the OEL. A statistical test of compatibility is used to determine whether the results from the various levels can be pooled, or whether a recovery curve needs to be employed. A target recovery for an acceptable method is typically 75% or greater. The study should use materials as close to those anticipated from the field as possible. Thus if the field samples are expected to contain refractory beryllium oxide, which is rather difficult to dissolve, a recovery study carried out only with soluble salts of beryllium may not be relevant.

Routine analyses should be carried out according to standard operating procedures. Investigations into poor quality performance have often uncovered deviations in procedure as the underlying cause. There are many internal quality control procedures that can be implemented in the chemical laboratory, and laboratories are also encouraged (sometimes required) to participate in external quality schemes such as round-robins, proficiency testing, and accreditation schemes. Proficiency testing is one way to determine analytical uncertainty. It can also be used to ensure the quality of on-site analyses. In OSHA parlance, the uncertainty of the sampling and analysis procedure is known as the sampling and analytical error (SAE). In determining compliance with a permissible exposure limit (PEL), OSHA subtracts the SAE from the measurement. Only if the measurement then exceeds the PEL is a citation likely.

2.11   Air Sampling in Retrospective Exposure Assessments

When it is necessary to try to recreate historical exposures does the paucity of sampling and analytical data become apparent. This is true for practically all industries and it is getting worse, as regulators start to recommend less quantitative decision criteria. It is possible, for example, to find a factory in full production for 30 years while there has been only a single campaign of area air samples on a single day. Years later, attempts will be made to reconstruct exposures using this as the only quantitative data available. Many beryllium workplaces, however, do not suffer from a similar lack of data. Table III of Kolanz *et al.*[195] includes facilities with thousands, even tens and hundreds of thousands, of samples, many of which were "low-flow personal lapel samples", although this dataset does also include large numbers of samples for which the analytical result was less than the limit of quantitation for the procedure, leading to issues of dealing with truncated datasets. The existence of these data nevertheless gives hope for exposure reconstructions that go well beyond what is available for most other industries. Nevertheless, exposure reconstructions still typically require some professional judgments involving assumption and extrapolation. Exposure data are then matched with detailed work histories to provide a useful job-exposure matrix.[210,211] The existence of a historic dataset, however, carries its own drawback; it leads to reluctance to update methodologies for fear of generating separate databases that cannot be linked. However, this should not be a barrier to innovation. It has been clearly shown that the variability inherent to worker-to-worker and day-to-day exposures typically far outweighs the variability in sampling and analytical methods.[22]

2.12   Conclusion

In an ideal world, the basis of a sampling method for an airborne particulate toxin would first involve a clear understanding of the mechanism of disease. This would allow the development of sampling methods that specifically target the particles of most concern. The consequence of not getting this right might be a breakdown in the dose–response equation if the most appropriate dose is not being measured. While measurements of numbers of fine and coarse particles tend to be linearly correlated, measurements of the masses of these particles tend to be exponentially correlated. Particle size-selective sampling has been around for many years, and the concept is not in dispute. There has been some dispute over the size-selective conventions and the calibration and suitability of equipment to measure them. No perfect sampler exists. Samplers may have drawbacks of suitability, *i.e.* there may be very high or very low aerosol concentration, high or low wind-speed, or directed wind, or there may be other environmental factors affecting the overall uncertainty of the method. There may be practical drawbacks of size, weight, cost, disposability and ease of calibration which limit the number of samples taken. The decision regarding the appropriate metric has swung over time from number to mass. It may be that the same metric should not be applied to all particles.

The current methodology to sample for airborne beryllium is the collection of airborne particles (aerosols) on filters held in holders. As described above, the 37 mm filter in its plastic, closed face (4 mm air entry inlet) cassette (CFC) holder has been the equipment of choice over the past 30–35 years in spite of many drawbacks (wall deposits, some handling issues, poor precision compared with some other samplers). This is also the current method of choice in the US for sampling many other airborne metals, including, for example, nickel and lead. The CFC has major practical advantages over other samplers with respect to being cheap and therefore disposable (ensuring no cross-contamination), lightweight and relatively unobtrusive, and easy to calibrate. An issue of importance is whether the sample represents a health-relevant fraction of the aerosol. When only the filter is analyzed from the CFC, it does not relate to any recognized size-selective collection efficiency for the evaluation of health effects. When the internal wall deposits of the CFC are included, the sample becomes much closer to the "inhalable" fraction as recognized by ISO and ACGIH®. It then becomes possible to argue over the small bias remaining and whether it is a rational trade-off for the practical advantages of the CFC over other samplers with less bias. Analysis of the wall deposits also is in line with OSHA procedures for metals, and OSHA compliance samples are analyzed in this way. However, the wall deposits of the CFC are not routinely analyzed by most other laboratories, so that one drawback for doing so is the likely increased results that will be obtained compared with historic data.

While the inhalable convention has been recommended for many metals, this does not necessarily make it the most appropriate convention for beryllium monitoring. A decision on the appropriate health-related fraction for beryllium should precede any discussion on sampling equipment selection. Finally, there is research underway to develop near real-time methods for beryllium detection. The sampling system in those methods should be subject to the same discussion with respect to health-relevance as when the analysis is end-shift or off-site.

Bibliography

The following text books are recommended for further information.

B. S. Cohen and C.S. McCammon, Jr, *Air Sampling Instruments* (9th edn), American Conference of Governmental Industrial Hygienists, Cincinnati, OH, 2001.

J. H. Ignacio and W.H. Bullock, *A Strategy for Assessing and Managing Occupational Exposures,* 3rd edn, American Industrial Hygiene Association Press, Fairfax, VA, 2006.

J. H. Vincent, Aerosol Sampling: Science, Standards, Instrumentation and Applications, J. Wiley, Chichester, UK, 2007.
K. Willeke and P. A. Baron, *Aerosol Measurement: Principles, Techniques and Applications,* 2nd edn, J. Wiley, Hoboken, NJ, 2005.

References

1. R. L. Naeye, *Chest*, 1973, **63**, 306–307.
2. J. H. Vincent, *J. Environ. Monit.*, 2005, **7**, 1037–1053.
3. T. Pearce, J. Hudnall, R. Lawrence, S. Martin, C. Coffey and J. Slaven, podium presentation #253 at the American Industrial Hygiene Conference & Exposition, Minneapolis, MN, 31 May–5 June 2008, www.aiha.org/abs08/08abindex.htm, accessed 10 February 2009.
4. W. H. Walton and J. H. Vincent, *Aerosol Sci. Technol.*, 1998, **28**, 417–438.
5. *2008 TLVs® and BEIs®. Based on the Documentation of the Threshold Limit Values for Chemical Substances and Physical Agents and Biological Exposure Indices*, American Conference of Governmental Industrial Hygienists, Cincinnati, OH, 2008.
6. *NIOSH Pocket Guide to Chemical Hazards*, National Institute for Occupational Safety and Health, Cincinnati, OH, 2005, Publ. 2005-151, www.cdc.gov/niosh/npg, accessed 10 February 2009.
7. US Code of Federal Regulations. Title 29 Part 1910 Subpart Z: Toxic and Hazardous Substances, US Occupational Health and Safety Administration, Washington, DC, www.osha.gov/pls/oshaweb/owastand.display_standard_group?p_toc_level=1&p_part_number=1910, accessed 10 February 2009.
8. Table 1: List of approved workplace exposure limits (as consolidated with amendments October 2007), in *EH 40/2005 Workplace Exposure Limits*, Health and Safety Executive, Sudbury, Suffolk, UK, www.hse.gov.uk/coshh/table1.pdf, accessed 10 February 2009.
9. S. M. Rappaport, *Am. J. Ind. Med.*, 1984, **6**, 291–303.
10. R. Tornero-Velez, E. Symanski, H. Kromhout, R. C. Yu and S. M. Rappaport, *Risk Anal.*, 1997, **17**, 279–292.
11. P. Hewett, *Appl. Occup. Environ. Hyg.*, 2001, **16**, 251–256.
12. Personal sampling for air contaminants, in *OSHA Technical Manual*, Occupational Safety and Health Administration, 2008, Section II, Chapter 1, www.osha.gov/dts/osta/otm/otm_ii/otm_ii_1.html, accessed 10 February 2009.
13. N. A. Leidel, K. A. Busch and J. R. Lynch (ed.), *Occupational Exposure Sampling Strategy Manual*, National Institute for Occupational Safety and Health, Cincinnati, OH, 1977, Publ. 77-173, www.cdc.gov/niosh/docs/77-173, accessed 10 February 2009.
14. EN 689:1996, *Workplace Atmospheres. Guidance for the Assessment of Exposure by Inhalation to Chemical Agents for Comparison with Limit Values and Measurement Strategy*, Comité Européen de Normalisation, Brussels, 1996.

15. S. M. Rappaport, R. H. Lyles and L. L. Kupper, *Ann. Occup. Hyg.*, 1995, **39**, 469–495.
16. J. C. Rock, in *Air Sampling for Evaluation of Atmospheric Contaminants, ed. B. S. Cohen and C. S. McCammon Jr*, American Conference of Governmental Industrial Hygienists, Cincinnati OH, 2001, ch. 2, pp. 20–50.
17. R. M. Tuggle, *Appl. Occup. Environ. Hyg.*, 2000, **15**, 380–386.
18. J. L. Hickey and P. C. Reist, *Am. Ind. Hyg. Assoc. J.*, 1977, **38**, 613–621.
19. S. A. Roach, *Am. Ind. Hyg. Assoc. J.*, 1978, **39**, 345–348.
20. R. S. Brief and R. A. Scala, *Am. Ind. Hyg. Assoc. J.*, 1986, **47**, 199–202.
21. J. Brodeur, A. Vyskocil, R. Tardif, G. Perrault, D. Drolet, G. Truchon and F. Lemay, *Am. Ind. Hyg. Assoc. J.*, 2001, **62**, 584–594.
22. H. Kromhout, E. Symanski and S. M. Rappaport, *Ann. Occup. Hyg.*, 1993, **37**, 253–270.
23. P. W. Logan and G. Ramachandran, in *A Strategy for Assessing and Managing Occupational Exposures*, ed. J. S. Ignacio and W. H. Bullock, American Industrial Hygiene Association, Fairfax, VA, 3rd edn, 2006, ch. 21, pp. 239–244.
24. D. Y. H. Pui, *Analyst*, 1996, **121**, 1215–1224.
25. J. H. Vincent, *Aerosol Sampling: Science, Standards, Instrumentation and Applications*, J. Wiley, Chichester, UK, ch. 20, 2007, pp. 489–515.
26. V. A. Marple and K. L. Rubow, *Am. Ind. Hyg. Assoc.*, 1978, **39**, 210–218.
27. P. A. Baron, *Analyst*, 1994, **119**, 35–40.
28. M. Linnainmaa, J. Laitinen, A. Leskinen, O. Sippula and P. Kalliokoski, *j. Occup. Environ. Hyg.*, 2008, **5**, 28–35.
29. D. Y. H. Pui. and D.-R. Chen, in *Air Sampling for Evaluation of Atmospheric Contaminants*, ed. B. S. Cohen and C. S. McCammon Jr, American Conference of Governmental Industrial Hygienists, Cincinnati, OH, 2001, ch. 15, pp. 377–414.
30. J. G. Olin, G. J. Sem and D. L. Christenson, *Am. Ind. Hyg. Assoc. J.*, 1971, **32**, 791–800.
31. H. Patashnick and G. Rupprecht, in *Proceedings of Advances in Particle Sampling and Measurement*, ed. W. B. Smith, Daytona Beach, FL, October 1979, US Environmental Protection Agency, 1980, Report EPA-600/9-80-004, p. 264.
32. P. Lilienfeld, *Am. Ind. Hyg. Assoc. J.*, 1970, **31**, 722–729.
33. J. C. Volkwein, E. Thimons, D. Dunham, H. Patashnick and E. Rupprecht, in *Proceedings of the 29th International Technical Conference on Coal Utilization and Fuel Systems*, Clearwater, FL, April 2004, Coal Technology Association, Gaithersburg, MD, vol. II, pp. 1355–1375.
34. J. T. Rozsa, J. Stone, O. W. Uguccini and R. E. Kupel, *Appl. Spectrosc.*, 1965, **19**, 7–10.
35. S. S. Cristy, presented at the Second Symposium on Beryllium Particulates and their Detection, Salt Lake City, UT, 8–9 November 2005, www.rmcoeh.utah.edu/besymp/presentations.pdf/Cristy%20Y12%20ATOFMS.pdf, accessed 10 February 2009.

36. M. Cheng, R. W. Smithwick III and R. Hinton, *J. ASTM Int.*, 2006, 3, DOI 10.1520/JAI13172.

37. S. Abeln, Y. Duan, J. A. Olivares, M. Koby and R. C. Scopsick, *Development of a Real-Time Monitor Utilizing Microwave Spectroscopy (MIPAES)*, Office of Science and Technical Information, US Department of Energy, 1999, Los Alamos National Laboratory Report LA-UR-99-3005 98531, www.osti.gov/bridge/servlets/purl/7591 978kKeac/webviewable/759197.PDF, accessed 10 February 2009.

38. G. E. Collins and G. Deng, *Metal Ion Analysis using Near-infrared Ddyes and the "Laboratory-on-a-Chip"*, 2003, report on Grant Number: DE-A107-98ER62711, #2, Office of Science and Technical Information, US Department. of Energy, www.osti.gov/em52/2003projsum/64982.pdf, accessed 10 February 2009.

39. B. S. Cohen, A. E. Chang, N. H. Harley and M. Lippmann, *Am. Ind. Hyg. Assoc. J.*, 1982, **43**, 239–243.

40. R. Vinson, J. Volkwein and L. McWilliams, *J. Occup. Environ. Hyg.*, 2007, **4**, 708–714.

41. J. Li, I. Yavuz, I. Celik and S. Guffey, *J. Occup. Environ. Hyg.*, 2007, **4**, 864–874. Erratum in *J. Occup. Environ. Hyg.*, 2007, **4**, D138.

42. B. S. Cohen, N. H. Harley and M. Lippmann, *Am. Ind. Hyg. Assoc. J.*, 1984, **45**, 187–192.

43. Method 7102, Beryllium and compounds (by graphite furnace atomic absorption), in *NIOSH Manual of Analytical Methods*, ed. P. C. Schlecht and P. F. O'Connor, National Institute for Occupational Safety and Health, Cincinnati, OH, 4th edn, 1994–2006, www.cdc.gov/niosh/nmam/, accessed 10 February 2009.

44. Method 7704 – Beryllium in air by field-portable fluorimetry, in *NIOSH Manual of Analytical Methods*, ed. P. C. Schlecht and P. F. O'Connor, National Institute for Occupational Safety and Health, Cincinnati, OH, 4th edn, 1994–2006, www.cdc.gov/niosh/nmam/, accessed 10 February 2009.

45. R. F. Phalen, in *Particle Size-Selective Sampling for Particulate Air Contaminants,* ed. J. H. Vincent, American Conference of Governmental Industrial Hygienists, Cincinnati, OH, 1999, pp. 29–49.

46. *Reference Manual*, Pergamon Press, Elmsford, NY, 1975, International Commission on Radiological Protection Publication 23.

47. G. Lidén and M. Harper, *J. Occup. Environ. Hyg.*, 2006, **3**, D94–D101.

48. *Identification and Description of Mineral Processing Sectors and Waste Streams: Beryllium*, US Environmental Protection Agency, Washington, DC, www.epa.gov/osw/nonhaz/industrial/special/mining/minedock/id/id4-ber.pdf.

49. D. Deubner, M. Kelsh, M. Shum, L. Maier, M. Kent and E. Lau, *Appl. Occup. Environ. Hyg.*, 2001, **16**, 579–592.

50. D. D. Thorat, T. N. Mahadevan and D. K. Ghosh, *Am. Ind. Hyg. Assoc. J.*, 2003, **64**, 522–527.

51. J. W. Martyny, M. D. Hoover, M. M. Mroz, K. Ellis, M. A. Maier, K. L. Sheff and L. S. Newman, *J. Occup. Environ. Med.*, 2000, **42**, 8–18.

52. Y. Thomassen, D. G. Ellingsen, K. Dahl, I. Martinsen, N. P. Skaugset and P. A. Drabløs, presented at Conférence Internationale de la Recherche sur le Béryllium, 8–11 March 2005. Montréal, QC, www.irsst.qc.ca/files/documents/PubIRSST/Be-2005/Session-8/Thomassen.pdf, accessed 10 February 2009.

53. O. A. Taiwo, presented at 3rd International Conference on Beryllium Disease, 16–19 October 2007, Philadelphia, PA, http://www.international beconference07.com, accessed 10 February 2009.

54. L. Godderis, W. Vanderheyden, J. Van Geel, G. Moens, R. Masschalein and H. Veulemans, *J. Environ. Monit.*, 2005, **7**, 1359–1363.

55. H. M. Cole, in *Managing Health in the Aluminium Industry*, proceedings of the International Conference on Managing Health Issues in the Aluminium Industry, 26–29 October 1997, Montréal, QC, ed. N. D. Priest and T. V. O'Donnell, Middlesex University Press, London, ch. 3, pp. 21–30, www.world-aluminium.org/cache/fl0000116.pdf, accessed 10 February 2009.

56. V. Bencko, E. V. Vasilieva and K. Symon, *Environ. Res.*, 1980, **22**, 439–449.

57. K. G. Crouch, A. S. Echt, R. Kurimo and Y. T. Gagnon, *Control technology and exposure assessment for occupational exposure to beryllium: abrasive blasting with coal-slag, National Institute for Occupational Safety and Health, Cincinnati*, OH, File No. EPHB 263-13a, 2007, www.cdc.gov/niosh/surveyreports/pdfs/ECTB-263-13a.pdf, accessed 10 February 2009.

58. L. D. Cunningham, *Beryllium Recycling in the United States in 2000*, US Department of the Interior, US Geological Survey, Reston, VA, USGA Circular 1196-P, 2004, http://pubs.usgs.gov/circ/c1196p/c1196p.pdf, accessed 10 February 2009.

59. *Frequently Asked Questions about Recycling, Disposal and Waste Classification of Beryllium*, Brush Wellman Corporation, FAQ205, 2006, www.brushwellman.com/EHS/FAQ/FAQ%20205.pdf, accessed 10 February 2009.

60. A. C. Fieldner, *The Sugar-Tube Method of Determining Rock Dust in Air*, US Department of the Interior, Bureau of Mines, Washington, DC, 1921, p. 278.

61. L. Greenburg, *A New Instrument for Sampling Aerial Dust. Report of Investigations*, US Department of the Interior, Bureau of Mines, Washington, DC, 1922, p. 2392.

62. R. J. Sherwood, *Appl. Occup. Environ. Hyg.*, 1997, **12**, 229–234.

63. A. L. Linch and M. Corn, *Am. Ind. Hyg. Assoc. J.*, 1965, **26**, 601–610.

64. C. E. Brown, *Filter-Paper Method for Obtaining Dust-Concentration Results Comparable to Impinger Results. Report of Investigations*, US Department of the Interior, Bureau of Mines, Washington, DC, 1944, p. 3788.

65. H. J. Paulus, N. A. Talvitie, D. A. Fraser and R. G. Keenan, *Am. Ind. Hyg. Assoc. Q.*, 1957, **18**, 267–273.

66. *US Federal Standard 209, Cleanroom and Work Station Requirements, Controlled Environments*. Available from Institute of Environmental Sciences and Technology, 940 East Northwest Highway, Mount Prospect,

IL 60056, USA, 1963 and later. (Withdrawn in 2001 but still widely used.)

67. *Air Sampling Instruments*, American Conference of Governmental Industrial Hygienists, Akron, OH, 1st edn, 1960.

68. Method S349, in *NIOSH Manual of Analytical Methods*, National Institute for Occupational Safety and Health, Cincinnati, OH, 2nd edn, 1977, vol. 3, Publ. 77-157-C.

69. *Documentation of the NIOSH Validation Tests, S262 and S349*, National Institute for Occupational Safety and Health, Cincinnati, OH, 1977, Publ. 77-185.

70. H. J. Beaulieu, A. V. Fidino, L. B. Kim, M. S. Arlington and R. M. Buchan, *Am. Ind. Hyg. Assoc. J.*, 1980, **41**, 758–765.

71. R. M. Buchan, S. C. Soderholm and M. I. Tillery, *Am. Ind. Hyg. Assoc. J.*, 1986, **47**, 825–831.

72. S. C. Soderholm, *Ann. Occup. Hyg.*, 1989, **33**, 301–320.

73. *ISO TR 7708-1983, Air Quality. Particle Size Fraction Definitions for Health-related Sampling*, International Organization for Standardization, Geneva, 1983.

74. L. D. Taylor, Letter to the editor, *Am. Ind. Hyg. Assoc. J.*, 1998, **59**, 8.

75. T. L. Ogden and J. L. Birkett, in *Inhaled Particles IV*, ed. W. H. Walton, Pergamon Press, Oxford, UK, 1977, pp. 93–105.

76. L. Armbruster and H. Breuer, in *Inhaled Particles V*, ed. W. H. Walton, Pergamon Press, Oxford, UK, 1982, pp. 21–32.

77. J. H. Vincent and D. Mark, in *Inhaled Particles V*, ed. W. H. Walton, Pergamon Press, Oxford, UK, 1982, pp. 3–19.

78. J. H. Vincent, D. Mark, B. G. Miller, L. Armbruster and T. L. Ogden, *J. Aerosol Sci.*, 1990, **21**, 577–586.

79. P. E. J. Baldwin and A. D. Maynard, *Ann. Occup. Hyg.*, 1998, **42**, 303–314.

80. R. J. Aitken, P.E.J. Baldwin, G. C. Beaumont, L. C. Kenny and A. D. Maynard, *J. Aerosol Sci.*, 1999, **30**, 613–626.

81. L. C. Kenny, R. J. Aitken, P. E. J. Baldwin, G. Beaumont and A. D. Maynard, *J. Aerosol Sci.*, 1999, **30**, 627–638.

82. O. Witschger, S. A. Grinshpun, S. Fauvel and G. Basso, *Ann. Occup. Hyg.*, 2004, **4**, 351–368.

83. D. Mark and J. H. Vincent, *Ann. Occup. Hyg.*, 1986, **30**, 89–102.

84. *EN 13205:2001, Workplace Atmospheres. Assessment of Performance of Instruments for Measurement of Airborne Particle Concentrations*, Comité Européen de Normalisation, Brussels, 2001.

85. L. C. Kenny, R. Aitken, C. Chalmers, J. F. Fabriès, E. Gonzalez-Fernandez, H. Kromhout, G. Lidén, D. Mark, G. Riediger and V. Prodi, *Ann. Occup. Hyg.*, 1997, **41**, 135–153.

86. M. S. Kent, T. G. Robbins and A. K. Madl, *Appl. Occup. Environ. Hyg.*, 2001, **16**, 539–558.

87. M. A. Werner, T. M. Spear and J. H. Vincent, *Analyst*, 1996, **121**, 1207–1214.

88. B. R. Conard, *Am. Ind. Hyg. Assoc. J.*, 1997, **58**, 629–630.

89. L. Brosseau, *Am. Ind. Hyg. Assoc. J.*, 1997, **58**, 631–632.
90. G. Wallis, Letter to the editor, *Am. Ind. Hyg. Assoc. J.*, 1998, **59**, 221–222.
91. *Criteria for a Recommended Standard: Occupational Exposure to Metal-working Fluids*, National Institute for Occupational Safety and Health, Cincinnati, OH, 1998, Publ. 98-102.
92. *Documentation of the Threshold Limit Values and Biological Exposure Indices, 7th edn,. Sulfuric Acid*, American Conference of Governmental Industrial Hygienists, Cincinnati, OH, 2004.
93. S. R. Woskie, T. J. Smith, M. F. Hallock, S. K. Hammond, F. Rosenthal, E. A. Eisen, D. Kriebel and I. A. Greaves, *Am. Ind. Hyg. Assoc. J.*, 1994, **55**, 20–29.
94. G. M. Piacitelli, W. K. Sieber, D. M. O'Brien, R. T. Hughes, R. A. Glaser and J. D. Catalano, *Am. Ind. Hyg. Assoc. J.*, 2001, **62**, 356–370.
95. H. M. Donaldson, R. A. Hiser and C. W. Schwenzfeier, *Am. Ind. Hyg. Assoc. J.*, 1964, **25**, 69–76.
96. A. J. Orenstein, in *Proceedings of the Pneumoconiosis Conference, Johannesburg, ed. A. J. Orenstein*, J. & A. Churchill Ltd, London, 1960.
97. T. F. Hatch and P. Gross, *Pulmonary Deposition and Retention of Inhaled Aerosols*, Academic Press, New York, 1964.
98. M. Lippmann, in *Air Sampling for Evaluation of Atmospheric Contaminants, ed.* B. S. Cohen and C. S. McCammon Jr, American Conference of Governmental Industrial Hygienists, Cincinnati, OH, 2001, ch 5, pp. 93–134.
99. L. C. Kenny and T. L. Ogden, *Ann. Occup. Hyg.*, 2000, **44**, 561–563.
100. G. A. Day, A. Dufresne, A. B. Stefaniak, C. R. Schuler, M. L. Stanton, W. E. Miller, M. S. Kent, D. C. Deubner, K. Kreiss and M. D. Hoover, *Ann. Occup. Hyg.*, 2007, **51**, 67–80.
101. ISO/TR 27628:2007, *Workplace Atmospheres. Ultrafine, Nanoparticle and Nano-structured Aerosols. Inhalation Exposure Characterization and Assessment*, International Organization for Standardization, Geneva, 2007.
102. M. Lippmann, in *Handbook of Physiology Section 9: Reactions to Environmental Agents, ed* D. H. K. Lee, H. L. Falk and S. D. Murray, The American Physiological Society, Bethesda, MD, 1977.
103. G. A. Kuhlmey, B. Y. H. Liu and V. A. Marple, *Am. Ind. Hyg. Assoc. J.*, 1981, **42**, 790–795.
104. J. Keskinen, K. Pietarinen and M. Lehtimäki, *J. Aerosol. Sci.*, 1992, **23**, 353–360.
105. J. K. Agarwal and G. J. Sem, *J. Aerosol Sci.*, 1980, **11**, 343–357.
106. E. O. Knutson and K. T. Whitby, *J. Aerosol Sci.*, 1975, **6**, 443–451.
107. A. Ankilov, A. Baklanov, M. Colhoun, K.-H. Enderle, J. Gras, Yu. Julanov, D. Kaller, L. Lindner, A. A. Lushnikov, R. Mavliev, F. McGovern, T. C. O'Connor, J. Podzimek, O. Preining, G. P. Reischl, R. Rudolf, G. J. Sem, W. W. Szymanski, E. Tamm, A. E. Vrtala, P. E. Wagner, W. Winklmayr and V. Zagaynov, *Atmos. Res.*, 2002, **62**, 209–237.

108. D5337-07, *Standard Practice for the Calibration of Air Sampling Pumps*, ASTM International, West Conshohocken, PA, 2007.

109. P. A. Baron, *Appl. Occup. Environ. Hyg.*, 2002, **17**, 593–597.

110. A. D. Maynard and L. C. Kenny, *J. Aerosol Sci.*, 1995, **26**, 671–684.

111. P. Baron, M. Box, A. Echt and S. Shulman, presented at the American Society for Testing and Materials Symposium on Silica: Sampling and Analysis, Salt Lake City, UT, April 22–23, 2004.

112. US Code of Federal Regulations, Title 30, Part 74.6(d).

113. G. Lidén, *J. Aerosol Sci.*, 2000, **31**(suppl. I), 270–271.

114. J. L. Weeks, *Am. Ind. Hyg. Assoc. J.*, 1995, **56**, 328–332.

115. J. L. Weeks, *Am. J. Ind. Med.*, 2007, **20**, 141–147.

116. J. P. Strange, S.P. Nebash, A.C. McInnes and P.W. McConnaughey, US Patent 3686835, 29 August 1972.

117. S. P. Nebash, US Patent 3957469, 3 February 1975.

118. G. F. Fafaul, *Development of an Improved Tamper-proof Filter Cassette*, US Department of Energy, 1978, open file report (final) PB-294186.

119. B. Y. H. Liu, D. Y. H. Pui and K. L. Rubow, in *Aerosols in the Mining and Industrial Work Environment, vol. 3, Instrumentation, ed.* V. A. Marple and B. Y. H. Liu, Ann Arbor Science, Ann Arbor, MI, 1981, pp. 989–1038.

120. J. McLister, P. R. Stacey and G. Revell, *Round Robin Filter Weighing Exercise,* Health and Safety Laboratory, Sheffield, UK, 2001.

121. Method 5040. Diesel particulate matter (as elemental carbon), in *NIOSH Manual of Analytical Methods*, ed. P. C. Schlecht and P.F. O'Connor, National Institute for Occupational Safety and Health, Cincinnati, OH, 4th edn, 2003, issue 3, www.cdc.gov/niosh/nmam/, accessed 10 February 2009.

122. Method 7500. Silica, crystalline, by XRD (filter redeposition), in *NIOSH Manual of Analytical Methods*, ed. P.C. Schlecht and P.F. O'Connor, National Institute for Occupational Safety and Health, Cincinnati, OH, 4th edn, 2003, issue 4, www.cdc.gov/niosh/nmam/, accessed 10 February 2009.

123. K. Ashley, R. N. Andrews, L. Cavazos and M. Demange, *J. Anal. At. Spectrom.*, 2001, **16**, 1147–1153.

124. Method 7300. Elements by ICP, in *NIOSH Manual of Analytical Methods*, ed. P.C. Schlecht and P.F. O'Connor, National Institute for Occupational Safety and Health, Cincinnati, OH, 4th edn, 2003, issue 3, www.cdc.gov/niosh/nmam/, accessed 10 February 2009.

125. P. L. Lowry and M. I. Tillery, *Filter Weight Stability Evaluation*, Los Alamos Scientific Laboratory, Los Alamos, NM, 1979, Report LA-8061-MS.

126. P. Berg, V. Jaakmees and L. Bodin, *Appl. Occup. Environ. Hyg.*, 1999, **14**, 592–597.

127. Å. Näsman, G. Blomquist and J.-O. Levin, *J. Environ. Monit.*, 1999, **1**, 361–365.

128. J. S. Webber, A. G. Czuhanich and L. J. Carhart, *J. Occup. Environ. Hyg.*, 2007, **4**, 780–789.

129. Method 5506. Polynuclear aromatic hydrocarbons by HPLC, in *NIOSH Manual of Analytical Methods*, ed. P. C. Schlecht and P. F. O'Connor,

National Institute for Occupational Safety and Health, Cincinnati, OH, 4th edn, 1998, issue 3, www.cdc.gov/niosh/nmam/, accessed 10 February 2009.

130. P. A. Lawless and C. E. Rhodes, *J. Air & Waste Manage. Assoc.*, 1999, **49**, 1039–1049.

131. M. Harper and M. Demange, *J. Occup. Environ. Hyg.*, 2007, **4**, D81–D86.

132. D. Mark, *Ann. Occup. Hyg.*, 1990, **34**, 281–291.

133. *Methods for the Determination of Hazardous Substances. General Methods for Sampling and Gravimetric Analysis of Respirable and Inhalable Dust*, Health and Safety Executive, Sudbury, Suffolk, UK, 2000, MDHS14/3.

134. *Methods for the Determination of Hazardous Substances. Metals in Air by ICP-AES*, Health and Safety Executive, Sudbury, Suffolk, UK, 2006, MDHS 99.

135. ISO 15202-2, *Workplace Air. Determination of Metals and Metalloids in Airborne Particulate Matter by Inductively Coupled Plasma Atomic Emission Spectrometry – Part 2: Sample Preparation*, International Organization for Standardization, Geneva, 2001.

136. J. P. Smith, D. L. Bartley and E. Kennedy, *Am. Ind. Hyg. Assoc. J.*, 1998, **59**, 582–585.

137. S.-N. Li. and D. A. Lundgren, *Am. Ind. Hyg. Assoc. J.*, 1999, **60**, 235–236.

138. G. Lidén and G. Bergman, *Ann. Occup. Hyg.*, 2001, **45**, 241–252.

139. F. Roger, G. Lachappelle, J. F. Fabriès, P. Görner and A. Renoux, *J. Aerosol Sci.*, 1998, **29** (suppl. 1), S1133–S1134.

140. G. Lidén and L. C. Kenny, *Ann. Occup. Hyg.*, 1994, **38**, 373–384.

141. N. P. Vaughan, C. P. Chalmers and R. A. Botham, *Ann. Occup. Hyg.*, 1990, **34**, 553–573.

142. R. J. Aitken and R. Donaldson, *Large Particle and Wall Deposition Effects in Inhalable Samplers*, Health and Safety Executive, Sudbury, Suffolk, UK, 1996, HSE Contract Research Report No. 117/1996.

143. G. Lidén, L. Juringe and A. Gudmundsson, *J. Aerosol Sci.*, 2000, **31**, 199–219.

144. V. Aizenberg, K. Choe, S. A. Grinshpun, K. Willeke and P. A. Baron, *J. Aerosol. Sci.*, 2001, **32**, 779–793.

145. G. Lidén, B. Melin, A. Lidblom, K. Lindberg and J.-O. Norén, *Appl. Occup. Environ. Hyg.*, 2000, **15**, 263–276.

146. D. J. Sivulka, B. R. Conard, G. W. Hall and J. H. Vincent, *Regul. Toxicol. Pharmacol.*, 2007, **48**, 19–34.

147. S. Kalatoor, S. A. Grinshpun, K. Willeke and P. A. Baron, *Atmos. Environ.*, 1995, **29**, 1105–1112.

148. V. Aizenberg, S. A. Grinshpun, K. Willeke, J. Smith and P. A. Baron, *Am. Ind. Hyg. Assoc. J.*, 2000, **61**, 398–404.

149. V. Aizenberg, K. Choe, S. A. Grinshpun, K. Willeke and P. A. Baron, *J. Aerosol Sci.*, 2001, **32**, 779–793.

150. Geräte zur Probenahme der einatembaren Staubfraktion (E-Staub), in *BIA No. 3010, BIA-Arbeitsmappe Messung von Gefahrstoffen*, Berufsgenossenschaftliches Institut für Arbeitsschutz (BIA), Erich Schmidt Verlag, Bielefeld, Germany, 2001.

151. C.-C. Chen and P. A. Baron, *Am. Ind. Hyg. Assoc. J.*, 1996, **57**, 142–152.
152. S.-N. Li, D. A. Lundgren and D. Rovell-Rix, *Am. Ind. Hyg. Assoc. J.*, 2000, **61**, 506–516.
153. S. Hetland and Y. Thomassen, *poster presentation at Airmon: Modern Principles of Workplace Air Monitoring,* Geilo, Norway, 1993.
154. M. Demange, J. C. Gendre, B. Hervé-Bazin, B. Carton and A. Peltier, *Ann. Occup. Hyg.*, 1990, **34**, 399–403, updated with additional data by the authors, October 2007.
155. M. Harper, B. Pacolay, P. Hintz and M. E. Andrew, *J. Environ. Monit.*, 2006, **8**, 384–392.
156. M. Harper, B. Pacolay and M. E. Andrew, *J. Environ. Monit.*, 2005, **7**, 592–597.
157. Method PV2121. Gravimetric Determination, in *OSHA Manual of Analytical Methods*, Occupational Safety and Health Administration, Salt Lake Technical Center, Sandy, UT, http://www.osha.gov/dts/sltc/methods/partial/pv2121/pv2121.html, Accessed 20 March, 2009.
158. M. Harper and B. Pacolay, *J. Environ. Monit.*, 2006, **8**, 140–146.
159. Method ID215 (version 2): hexavalent chromium, in *OSHA Sampling and Analysis Methods*, Occupational Safety and Health Administration, Salt Lake City, UT, rev. 2006.
160. F. Stones, S. Edwards D. Crane and G. Schultz, poster presented at American Industrial Hygiene Conference & Exposition, Atlanta, GA, 8–13 May 2004.
161. *MétroPol, Fiche No. 003: Métaux – Metalloïdes.* Institut National de Recherche et Sécurité, Vandœuvre-les-Nancy, France, 2005.
162. M. A. Puskar, J. M. Harkins, J. D. Mooney and L. H. Heckler, *Am. Ind. Hyg. Assoc. J.*, 1991, **52**, 280–286.
163. M. Demange, P. Görner, J.-M. Elcabache and R. Wrobel, *Appl. Occup. Environ. Hyg.*, 2002, **17**, 200–208.
164. P. Görner, R. Wrobel, F. Roger and J.-F. Fabriès, *J. Aerosol Sci.*, 1999, **30**(suppl. 1), S893–S894.
165. M. Harper, *J. Environ. Monit.*, 2006, **8**, 598–604.
166. N. J. Kennedy and W. C. Hinds, *J. Aerosol Sci.*, 2002, **33**, 237–255.
167. D. J. Hsu and D. L. Swift, *J. Aerosol Sci.*, 1999, **30**, 1331–1343.
168. P. N. Breysse and D. L. Swift, *Aerosol Sci. Technol.*, 1990, **13**, 459–464.
169. Y. T. Dai, Y. J. Juang, Y. Wu, P. N. Breysse and D. J. Hsu, *J. Aerosol Sci.*, 2006, **37**, 967–973.
170. M. Harper, M. Z. Akbar and M. E. Andrew, *J. Environ. Monit.*, 2004, **6**, 18–22.
171. L. Dobson, L. Reichmann and D. Popp, *J. ASTM Intl.*, 2005, 2, DOI: JAI12229.
172. D. L. Bartley, C.-C. Chen, R. Song and T. J. Fischbach, *Am. Ind. Hyg. Assoc. J.*, 1994, **55**, 1036–1046.
173. P. Görner, R. Wrobel, V. Mièka, V. Škoda, J. Denis and J.-F. Fabriès, *Ann. Occup. Hyg.*, 2001, **45**, 43–54.

174. G. Lidén, *J. Aerosol Sci.*, 2000, **31**(suppl. 1), 270–271.

175. H. J. Ettinger, J. E. Partridge and G. W. Royer, *Am. Ind. Hyg. Assoc. J.*, 1970, **31**, 537–545.

176. J. C. Tsai and T. S. Shih, *Am. Ind. Hyg. Assoc. J.*, 1995, **56**, 911–918.

177. M. Gautam and A. Sreenath, *J. Aerosol Sci.*, 1998, **27**, 1265–1281.

178. C.-C. Chen and S. H. Huang, *Am. Ind. Hyg. Assoc. J.*, 1999, **60**, 720–729.

179. J. C. Tsai, H. G. Shiau, K. C. Lin and T. S. Shih, *J. Aerosol Sci.*, 1999, **30**, 313–323.

180. J. C. Tsai, H. G. Shiau and T. S. Shih, *Aerosol Sci. Technol.*, 1999, **31**, 463–472.

181. G. Lidén, *Appl. Occup. Environ. Hyg.*, 1993, **8**, 178–190.

182. M. Harper, C.-P. Fang, D. L. Bartley and B. S. Cohen, *J. Aerosol Sci.*, 1998, **29** (suppl. 1), S347–S348.

183. R. J. Higgins and P. Dewell, in *Inhaled Particles and Vapors II, ed. C. N. Davies,* Pergamon Press, Oxford, UK, 1967, pp. 575–586.

184. L. C. Kenny and R. A. Gussman, *J. Aerosol Sci.*, 1997, **28**, 677–688.

185. Geräte zur Probenahme der alveolengängige Staubfraktion (A-Staub), in *BIA No. 6068, BIA-Arbeitsmappe Messung von Gefahrstoffen,* Berufsgenossenschaftliches Institut für Arbeitsschutz, Erich Schmidt Verlag, Bielefeld, Germany, 2002.

186. L. C. Kenny, K. Chung, M. Dilworth, C. Hammond, J. Wynn Jones, Z. Shreeve and J. Winton, *Ann. Occup. Hyg.*, 2001, **45**, 35–42.

187. J. D. Stancliffe and L. C. Kenny, *J. Aerosol Sci.*, 1997, **28** (suppl. 1), S601–S602.

188. J. D. Stancliffe and K. Y. K. Chung, *Preparation, handling loading and analysis of PUF plugs used for welding fume sampling,* Health and Safety Laboratory, Buxton, UK, 1997, Report IR/L/A/97/09.

189. W. Koch, W. Dunkhorst and H. Lödding, *Aerosol Sci. Technol.*, 1999, **31**, 231–246.

190. R. Rando, H. Poovey, D. Mokadam, J. Brisolara and H. Glindmeyer, *J. Occup. Environ. Hyg.*, 2005, **2**, 219–226.

191. W. Koch, W. Dunkhorst, H. Lödding, Y. Thomassen, N. P. Skaugset, A. Nikanov and J. Vincent, *J. Environ. Monit.*, 2002, **4**, 657–662.

192. K. R. May, *J. Sci. Instrum.*, 1945, **22**, 187–197.

193. V. A. Marple and J. E. McCormack, *Am. Ind. Hyg. Assoc. J.*, 1983, **44**, 916–922.

194. H. Gibson, J. H. Vincent and D. Mark, *Ann. Occup. Hyg.*, 1987, **31**, 463–479.

195. M. E. Kolanz, A. K. Madl, M. A. Kelsh, M. S. Kent, R. M. Kalmes and D. J. Paustenbach, *Appl. Occup. Environ. Hyg.*, 2001, **16**, 593–614.

196. *Beryllium Air Sampling Procedure,* US Department of Energy, Office of Health, Safety and Security, www.hss.energy.gov/HealthSafety/WSHP/be/guide/itk/12690.pdf, accessed 10 February 2009.

197. H. M. Donaldson and W. T. Stringer, *Beryllium Sampling Methods: Comparison of Two Personal Sample Collection Methods with the AEC Sample Collection Method as Used for One year in a Beryllium Production*

Facility, National Institute for Occupational Safety and Health, Cincinnati, OH, 1976, Publ. 76-201.

198. P. C. Kelleher, J. W. Martyny, M. M. Mroz, L. A. Maier Ruttenber, D. A. Young and L. S. Newman, *J. Occup Environ Med.*, 2001, **43**, 238–249.

199. O. G. Raabe, *in Advances in Air Sampling, American Conference of Governmental Industrial Hygienists*, Lewis Publishers, Chelsea, MI, 1988, pp. 39–51.

200. E. King, A. Conchie, D. Hiett and B. Milligan, *Ann. Occup. Hyg.*, 1979, **22**, 213–239.

201. M. A. McCawley, M. S. Kent and M. T. Berakis, *Appl. Occup. Environ. Hyg.*, 2001, **16**, 631–638.

202. A. B. Stefaniak, M. D. Hoover, G. A. Day, R. M. Dickerson, E. J. Peterson, M. S. Kent, C. R. Schuler, P. N. Breysse and R. C. Scripsick, *J. Environ. Monit.*, 2004, **6**, 523–542.

203. US Environmental Protection Agency, Title 40 Part 61 Subpart C: *National Emission Standard for Beryllium*, 38 FR 8826, 6 April 1973, as amended at 65 FR 62151, 17 October 2000.

204. US Code of Federal Regulations Title 40 Part 60 Appendix A (revised February, 2000) Method 5. Determination of particulate matter emissions from stationary sources, US EPA Emissions Measurement Center, Research Triangle Park, NC, pp. 371–442, www.epa.gov/ttn/emc/promgate/m-05.pdf, accessed 10 February 2009.

205. US Code of Federal Regulations Title 40 Part 60 Appendix A (revised February, 2000) Method 104. Determination of beryllium emissions from stationary sources, US EPA Emissions Measurement Center, Research Triangle Park, NC, pp. 1776–1794, www.epa.gov/ttn/emc/promgate/m-104.pdf, accessed 10 February 2009.

206. US Code of Federal Regulations Title 40 Part 60 Appendix A (revised February, 2000) Method 103. Beryllium screening method, US EPA Emissions Measurement Center, Research Triangle Park, NC, pp. 1764–1775, www.epa.gov/ttn/emc/promgate/m-103.pdf, accessed 10 February 2009.

207. E. R. Kennedy, T. J. Fischbach, R. Song, P. M. Eller and S. A. Shulman, *Guidelines for Air Sampling and Analytical Method Development and Evaluation*, National Institute for Occupational Safety and Health, Cincinnati, OH, 1995, Publ. 95-117.

208. EN 482:1994, *Workplace Atmospheres. General Requirements for the Performance of Procedures for the Measurement of Chemical Agents*, Comité Européen de Normalisation, Brussels, 1994.

209. D. L. Bartley, J. E. Slaven, M. E. Andrew, M. C. Rose and M. Harper, *J. Occup. Environ. Hyg.*, 2007, **4**, 931–942.

210. A. B. Stefaniak, V. M. Weaver, M. Cadorette, L. G. Puckett, B. S. Schwartz, L. D. Wiggs, M. D. Jankowski and P. N. Breysse, *J. Occup. Environ. Hyg.*, 2003, **18**, 708–715.

211. A. E. Barnard, J. Torma-Krajewski and S. M. Viet, *Am. Ind. Hyg. Assoc. J.*, 1996, **57**, 804–808.

CHAPTER 3
Surface Sampling[*‡]
Successful Surface Sampling for Beryllium

GLENN L. RONDEAU

Environmental Safety & Health Contract Services Safety Specialist,
Department of Energy, National Energy Technology Laboratory, Albany,
Oregon 97321, USA

Abstract

Accurate surface sampling depends on comprehensive and precise pre-planning
that quantifies what will be accomplished, when it is to be done, how it is to be
done, and who will do it. Careful attention to detail and thorough documentation
at each step of the process is critical to achieving accurate and defensible results.
This chapter explores each of these important successful sampling elements.

3.1 Surface Sampling

At US Department of Energy (DOE) sites, surface sampling for beryllium
contamination is required by 10 CFR 850.20.b.4, which states that the

---

[*]This article was prepared by a US Government contractor employee as part of his official duties.
The US Government retains a nonexclusive, paid-up, irrevocable license to publish or reproduce
this work, or allow others to do so for US Government purposes.
[‡]*Disclaimer:* The findings and conclusions in this chapter are those of the author and do not
necessarily represent the views of the US Department of Energy. Mention of sampling devices or
software does not constitute an endorsement, and does not imply that other devices are not fit for
the same purpose.

Beryllium: Environmental Analysis and Monitoring
Edited by Michael J. Brisson and Amy A. Ekechukwu
© Royal Society of Chemistry 2009
Published by the Royal Society of Chemistry, www.rsc.org

responsible employer "in conducting the baseline inventory, must conduct air, surface, and bulk sampling."[1,2] Surface sampling is required to determine if beryllium contamination may exist, and if it does, at what levels and where.[1–5] For legacy areas, surface sampling provides the evidence used to determine that remediation is or is not necessary.

Surface sampling is accomplished by one of the following methods: wipe sampling; bulk sampling; and vacuum sampling. Sampling methods to be used depend largely on the conditions and type of surfaces to be sampled. The following is a partial listing of variables to be considered in selecting surface sampling technique(s):

- Surface clean, dusty or dirty, smooth or rough or porous
- Wipe *vs.* bulk
- Dry wipe *vs.* wet wipe *vs.* solvent or alcohol wipe
- Bulk sampling sweep, vacuum or particle mass

3.1.1   Wipe Sampling

Wipe sampling consists of dry or agent moistened (wet) media applied in a defined and deliberate method of motion and direction, with consistently applied pressure of the wiping media onto a defined surface area. At DOE sites, results are reported in µg beryllium per $100\,cm^2$. Some wetting agents are distilled water, alcohol, proprietary solution on packaged wipes, mild detergents (be careful with these due to possible background interferences) and solvents (acetone, toluene, *etc.*).[6] Wipe sampling should be done on relatively clean surfaces per regulatory guidance.[1,2] Effective wet wipe sampling produces the greatest recovery of beryllium if there are clean smooth surfaces.[7–10] Wet wiping should be used unless wetting the surface would change the properties of the wiped surface,[8] or the quantity of dust requires the use of bulk sampling.[8,11]

Wear clean, impervious, disposable powderless gloves when taking wipe samples to prevent sample contamination. Change gloves after placement of template(s) before taking a sample, and also between samples to reduce the possibility of cross-contamination.[12] Determine if dry or wet wipe sampling will be done, and whether it is to be with a pre-moistened Ghost Wipe™, an alcohol or solvent wetted pad for oily surface, or a dry filter such as a Whatman® filter paper or other medium. Follow the folding sequence of removing the medium from its package wearing the gloves, unfold the pad, and then fold it in half and then fold in half again. You are now ready to begin the wiping sequence by wiping in a Z or S pattern one direction across the area to be sampled. Wipe slowly and apply even pressure during this and all wiping sequences. Next, fold the wiping pad in half so that the exposed wiped portion of the pad is folded in to expose a clean pad surface. Using the clean portion of the pad, wipe the sampled area with the same Z or S pattern wiping at a right angle to the previous series across the sampled area. When through, fold this exposed wiped portion into the pad to expose a clean portion. With this, wipe the sampled area again horizontally and vertically around the edges of the sampled area. Standard methods utilizing the S pattern have been promulgated by ASTM International for both wet and dry wipes.[13,14] Note that some pre-packaged pre-moistened wipes may be too wet for wiping a cleaned or small area; in this case, squeeze some of the liquid out before beginning the wipe sequence to avoid leaving puddles of liquid on the wiped surface. Collect the liquid if this occurs.

When through, carefully place the exposed wiping pad in a labeled container, such as a 50 ml digestion vial, and label and seal it for sending to the laboratory for analysis.[12] Be certain to complete the sampling and oversight recordkeeping and take a photo of the sample location.[15] Occasionally it is also necessary to include field blanks in the batch of samples being sent to the laboratory, but not more than the spiked or reference material samples. It is important to follow a chain-of-custody method to document involvements from the point of collecting the sample to its final disposition.

A wipe sample of skin could be taken with a pre-moistened Ghost Wipe™, a dry filter or a swab to determine and quantify the extent of possible exposure. Workers handling beryllium containing or contaminated items or processes should be wearing appropriate (and required) personal protective equipment (PPE) and working within protective processes and procedures.[1–5,16] Nevertheless, skin exposures can occur as result of improperly wearing or removing PPE or touching one's face with a contaminated glove, *etc.*[17] Surface sampling of a worker's skin or work clothing may be indicated for workers in dusty environments as skin is one route of exposure entry to be protected against.[16,18–19] Work clothing sampling can be done with vacuum methods or by having a piece of the clothing itself analyzed. Check with the laboratory to determine if they can analyze the clothing and what they need.

3.1.2 Bulk Sampling

Bulk sampling consists of collecting mass by brushing, sweeping, vacuuming (with use of filter cassette or HEPA sock collection devices) for dusts and larger sized particles. Analysis results are reported as parts per million (ppm) or microgram beryllium per gram ($\mu g\,g^{-1}$). Bulk sampling is done on dirty or heavily loaded surfaces,[8,11] or to determine beryllium content of materials that might present contamination if disturbed.[1,20] Soil and other naturally occurring materials, for example, contain beryllium and it is necessary to distinguish between these surfaces and other anthropogenic beryllium sources.[6,20]

In order of laboratory and/or the sampling plan preference, bulk samples may be one of the following:[12]

- A high-volume vacuum filter sample
- A representative sweeping of settled dust (*i.e.* rafter) sample
- A sample of the bulk material (or soil) in the workplace
- A sample taken with a wipe

Vacuum samples can be taken as described for wipe samples moving the vacuum nozzle in a slow and even pressure Z or S pattern across the designated sample area to obtain an analysis result of µg beryllium per 100 cm^2 of surface area.[21,22] If only interested in amount of µg beryllium per gram or ppm of beryllium in the sample, the specific Z or S sequence is not needed. Either a pre-weighed cassette filter or a pre-weighed HEPA filter vacuum bag[23] should be used. Be certain to obtain an adequate amount of sample for the laboratory to analyze. Bulk samples should be separated from other samples when shipping to the laboratory.[12]

For shipping the sample to the laboratory, label and seal the vacuum cassette or HEPA filter bag, or transfer the bulk material in a labeled container sealing it to prevent tampering and send to the laboratory for analysis.[12] Occasionally, send a field blank with the batch of samples to be analyzed. It is important to follow a chain-of-custody protocol to document involvements.

3.1.3 Vacuum Sampling

High volume vacuum or micro-vacuum sampling can be used to complement other surface sampling techniques such as wipe sampling, but this method can have a widely fluctuating performance depending on the surface vacuumed. Collection extractions can range from 25% to 50% for smooth surfaces such as glass, tile, steel, linoleum, vinyl, and wood; and from 21% to 85% for rough or porous surfaces such as wood, cloth, carpet, and concrete block.[24,25] On rough surfaces, it can be impractical to wipe sample,[24–26] but be aware that loose substrate particles will be picked up with the vacuumed dust, so it is important to know what this material is or what it could be. It might be an interference substance that could produce a false report of beryllium when analyzed.[15,27–29] It is important to homogenize the sample, if possible, before sending to the laboratory for analysis.[28]

3.2 Locations of Sample Points and Number of Samples

3.2.1 Randomly Selected Sample Points

Randomly Selected Sample Points are determined by various methods. Software, such as Visual Sampling Plan (VSP),[30] can be used to randomly designate the locations to be sampled within the sampling space. The necessary numbers of samples can be dependent upon sample results, with sampling being done on a sequential sampling approach. This can begin with a few samples and taking more as needed,[31] up to maximum of 59 samples[30] for a given homogeneous area. Care should be taken to avoid hyper geometric or over sampling. A homogeneous area would be an area or equipment with similar history and use as concerning potential for beryllium contamination. This could be a unique room or equipment, or it could be a grouping of offices, restrooms, production area, light fixtures, *etc.*[17,32] It could also be stratified or layered areas such as floors *vs.* walls *vs.* ceilings.[15] Randomly designated sample points could be assigned within a stratified area, or the area could be screened using biased or judgmental sampling.

3.2.2   Biased or Judgmental Sample Points

Biased or judgmental sample points are not determined randomly, but rather are based on a history of use and sampler judgment specific to criteria, such as beryllium was worked on a particular machine or in area in a room. A sample is purposely taken at such locations, since that is where contamination is most likely to be found. For example, if the spread of contamination is suspected to be the result of tracking from area to area on shoes or mobile equipment such as carts, fork lifts, *etc.*, biased or judgmental sampling would be done on floors, especially at entrance/egress points of areas where beryllium dusts are known or suspected to exist. People stop at, shuffle or walk through doorways with heavier steps than when traversing an area and, therefore, these areas have a high potential for beryllium to be deposited on the surface tile, cement, or carpet surfaces.

3.3   Sampling Techniques

3.3.1   Speed and Pressure

The speed and pressure of taking the sample are critical to obtaining correct results when taking a wipe or bulk sample. Samplers and contractors prefer wipe sampling over bulk sampling because wipe sampling is easier and takes less time, as stated previously. However, going too fast, especially with wipes, can result in not collecting all of the surface material. This can occur if the wet wipe agent does not have enough residence time on the surface to loosen the surface material so that all of it is collected.[15] The amount of hand pressure is also critical to collecting a complete sample from the surface.[12] See Table 3.1 for technique factors which, if applied incorrectly, can produce sampling errors with resultant useless or misleading data.[12,15,17,24,25]

3.3.2   Selection of Sampling Medium

Selection of an appropriate wipe sampling medium or technique is a question that is usually resolved in the field by the sampler at the time of taking the specific sample. However, this decision should be based on guidance provided in the sampling plan, *i.e.* specifics on what media or technique to be used when confronted with what situation such as oily surfaces, clean *vs.* dusty surfaces, rough *vs.* smooth surfaces, *etc.*

Below are some media choices for performing wipe sampling:[6]

- Ghost Wipes™ (PVA) pre-moistened or dry
- Lead Wipes™

Table 3.1 Sampling technique factors that compromise sampling.

- Not following the standard
- Not following the sampling plan
- Contamination of the sampling media
- Contamination of the obtained sample
- Alteration (deliberate or inadvertent) of to-be-sampled surface dust
- Applying uneven pressure against the surface when sampling
- Reaching rather than placing oneself at or over the sampling point
- Going too fast (inadequate residence or contact time with surface)
- Incorrect technique or media:
  - Using wipe when should use bulk
  - Using pre-packaged watered wet wipe on oily surface
  - Using dry wipe when wetted wipe should be used
  - Using inadequate technique (sweep vs. vacuum on rough surfaces)
- Inappropriate location specified for sample to be taken
- Not taking sample at specified location for reasons of convenience or fear
- Sampling at heights (ceilings, walls using ladders, man lifts)
- Sampling in congested tight quarters (under desks, in confined spaces, sampling in awkward position vs repositioning)
- Not documenting sample taken was relocated from specified location
- Not documenting that sample was taken on differing types of surfaces
- Incorrect measurements of surface sample area:
  - Smaller than actual can produce high results
  - Larger than actual can produce lower results
- Photos not correct, not cross referenced correctly
- Time constraints influencing incorrect decisions on sampling by samplers
- End of shift, lunch time, bath room breaks resulting in:
  - Hurried sampling, incomplete sampling or recordkeeping
  - Mistakes made but not corrected, covered up
  - Not sampled, faked sampling
- Personal factors – phone calls, health, other
- Heat stress, fatigue causing sampler to lose judgment or control

- Palintest® Dust Wipes™
- Linen circles (*e.g.* Nucon smears)
- Ashless filter papers (*e.g.* Whatman® 541 or similar filter paper)
- Pace Wipes™
- Smear Tabs™(paper)
- Nextteq® Microteq® Beryllium ChemTest™ colorimetric wipe

3.3.3 Determining Surface Area

The surface area must be known and communicated to the laboratory on the labels of wipe samples in order for it to calculate and report the analyzed result as µg beryllium per 100 cm^2. The sample area can be measured for the specific sample or a template can be used. When wipe sampling, the minimum area to sample should be 100 cm^2. If it is necessary to wipe a smaller area, wipe the entire object surfaces.[1,2] Commercially available 100 cm^2 templates can be used;

however, these and smaller surface areas introduce the possibility that the results may be censored or less than the laboratory reporting limit. Censored data pose problems when evaluating the collective data when the sample area is small.[31] Larger sized areas and commercially available templates, such as the 512 cm[2] hard stock paper frames used for mounting overhead transparencies, are recommended.[15] Larger sample areas help to avoid censored data concerns by providing a larger sample size that provides for more accurate results of analysis.[31]

3.3.4 Field Analysis

Field analysis of samples is desired to provide the ability to make decisions while sampling, especially when performing exposure or biased/judgmental sampling. The quantitative method using dilute ammonium bifluoride extraction and optical fluorescence has been adopted as an ASTM *Standard Test Method for Determination of Beryllium in Soil, Rock, Sediment and Fly Ash.*[33,34] A less expensive semi-quantitative colorimetric wipe method that enables field sample analysis is the Nextteq® Microteq® Beryllium ChemTest™, which can detect beryllium as low as 0.2 μg depending on the presence of any interferences, with detection being the resultant color changes of the wipe.[35] The 0.2 μg detection limit is the present DOE surface action level,[1,2] and the resultant color change interpretation is subjective by the analyst.[35] Recognizing and working within these limitations, this semi-quantitative colorimetric wipe method can be useful in screening and other applications where quick results are needed. Applications can include assessing effectiveness of control procedures, cleaning and decontamination, PPE effectiveness or possible breakthroughs, release of equipment, or early characterization sampling of areas to determine the need for more extensive sampling.[35–38]

3.3.5 Protecting Sample Process and Samples from Contamination

Sampling often involves situations where it is easy to contaminate the sample in the process. Samplers need to be alert to this and to take precautions so that it does not occur. For example, when placing a template and then taking a sample, stop your gloved hands from transferring contamination to the sampling template or sample. To prevent this, wear two or more gloves, remove the top glove that placed the template, and then sample with the clean remaining glove. This practice is also used when climbing a ladder to place a template and sample (in this case, triple gloving is often used). This technique emphasizes the importance of purchasing good quality and appropriately sized disposable gloves, so that the gloves do not rip apart when wearing multiple layered gloves. Powderless nitrile gloves are a must. Talc, which may be the powder used, can contain beryllium. To further help prevent contamination, samplers and recorders may utilize carpenters' aprons with pockets to keep necessary supplies available and clean. In some cases, it is advisable to have these supplies only at the cart or in a bucket. It is sometimes easier to prevent contamination when obtaining a bulk sample than a wipe, but it is equally critical not to contaminate the sample. If the sample has been potentially contaminated, this should be noted on recordkeeping and oversight documents for evaluation and decision-making about the sample.[15]

3.3.6 Inappropriate Sampling or Techniques

Samplers may tend to perform wet wipe sampling on irregular surfaces that do not lend themselves to good recovery by wipe sampling. There are various reasons for this, not the least of which is that wet wipe sampling with a premoistened pre-packaged wipe with a template is easier to do and takes less time than other alternative techniques such as vacuuming. Vacuum sampling, for example, allows for good recovery on surfaces where wipe sampling is impractical. Wipe sampling is not recommended for rough and/or porous surfaces, such as cinder block or carpet, or fragile substrates such as aged and deteriorated painted surfaces. A viable alternative to be considered is sampling by suction techniques.[15,21,22,24–26]

3.4 Sample Planning

Getting organized is essential to avoiding surprises and delays, whether the sampling to be done is an individual one-time sample to determine if a piece of equipment is contaminated, or a large, complex sampling campaign to characterize a site.

3.4.1 Determine Needs

Specify the sampling need or goal.[28,39] Is it to characterize or assess the extent, if any, legacy issues of potential beryllium contamination of an area or building or room? Or, is it simply to determine if a piece of equipment can be removed from a previously determined or suspected to be contaminated room without special handling. These are very different needs that require different approaches even though both require sampling and both require planning. Within each of these are variations such as: [1–4,6,8,11,28,31,32,40]

- Housekeeping: routine or incident response
- Control assessment: routine or incident response
- Facility assessment: scoping, evaluation, clearance/release
- Equipment release: batches/streams of similar items or single item
- Bulk sampling: soils comparison, material beryllium content, waste characterization, remediation effectiveness

3.4.2   Contaminated Surfaces

Contaminated surfaces can include equipment, tools, floors and stairs, work clothing and areas of exposed skin on workers, visitors, and samplers.[5] Before starting to sample, the sampling campaign's objective must be defined so that the who, when, how, *etc.* can be specified and planned for.[17,41] The recommended number and location of wipe samples can be found by reviewing sampling plans and procedures in documents in the *Chronic Beryllium Disease Prevention Program (CBDPP) Implementation Tool Kit.*[2]

3.4.3   Planning Tools

Sampling campaign type, scope and strategy, and statistical tools such as EPA's ProUCL 4.0,[42] Upper Tolerance Level UTL,[43] MARSSIM,[44,45] Visual Sampling Plan (VSP),[30] and The R Program[46] are very helpful in the sampling process.[8,45] They assist with the number of samples, population, the randomization of sample points, and subsequent evaluation of the analytical results.[10] VSP,[30] for example, randomly selects locations to be sampled and graphically plots out these selected sample locations in three-dimensional (3-D) space, as well as evaluating the subsequent results and drafting a final conclusion report.[8,10,15,30] Statistical programs such as these are essential in surface sampling projects involving the characterization of facilities, and should be an integral part of any sampling campaign. For example, VSP can be used for randomization and other tools can be used for deciding on the number of samples and choice of statistical methods. These statistical tools are also utilized to evaluate sample data.[15,28,46] In some cases, very few samples are needed to characterize the area as clean.[31]

3.4.4   Standard Operating Procedure

A Standard Operating Procedure (SOP) or a standardized form can easily cover the need for routine day-to-day sampling, such as movement or release of equipment, to be done consistently in a way that records are produced for defensibility concerns. Complete recordkeeping is essential.[1–4,6,15] On sampling campaigns, a detailed plan, specific to needs and goals, is required with responsible parties agreeing to follow the plan as written. To ensure defensibility of the effort, a well-documented oversight element is required to capture what occurred, who did it, how it was done, and any problems that might have influenced the given sample results.

3.4.5   Overall Sampling Plan

The overall sampling plan must include a detailed safety plan to ensure protection of the personnel involved in the sampling effort as well as the building occupants, visitors, and the environment. Issues such as sensitive equipment, hazards, working hours, and other elements need to be documented and communicated. See Table 3.2 for examples that are included in a typical Safety Plan.[1–5,11,18,19,27]

Table 3.2 Sampling safety plan considerations.[5,23]

Topic to be considered and documented in plan	Who is affected or covered		
	Samplers	*Occupant*	*Visitors*
• Insurance (include copy of policy in plan)			
• Qualifications of samplers	X		
• Driver's licenses, certifications	X		
• Staff medical clearances	X	X	
• Ability to use respirators	X	X	
• Beryllium Lymphocyte Proliferation Test (BeLPT)[32]	X	X	
• Medical restrictions	X	X	X
• Special medical needs	X	X	
• Staff availability (known absences identified)	X	X	
• Staff emergency contact information	X	X	
○ Where staying – hotel, home	X	X	
○ Who to notify in event of emergency	X	X	
■ At the site	X		
■ Family member contact	X	X	
• Hours to be worked daily on the project	X	X	
• Scheduled breaks and lunch times for each day	X		
• How will meals be provided (brought in or go out)	X	X	X
• Where will breaks or lunches be taken?	X	X	X
• Rest room accommodations (use host's or portable?)	X	X	X
• Showering needs (use host's or portable?)	X		
• Pre-requisite training	X	X	
• Orientation training to do at start of project	X		
• Supplies, equipment to be brought to project	X		
• Supplies to be furnished during project and by whom	X	X	
• Equipment to be furnished during project and by whom:	X		
○ Personal protection equipment	X		
○ Ladders, aerial lift, fork truck		X	X
○ Waste bags, barrier tape, batteries, *etc.*	X	X	X
• Sampling safety:	X	X	
○ Protection of sampler	X	X	X
○ Protection of sampler's work area	X	X	X
○ Protection of area occupants and their area	X	X	X
○ Protection of environment	X	X	
• How will sampling and food wastes be handled	X	X	
• Safety oversight – who will do what, when and how	X	X	
• Emergency contact information:			
○ Procedures to be followed			
○ Samplers' employer information			
○ Samplers' family contacts			
○ Samplers' hotel information			

3.4.6 Site History

Site history is important to guide both the sampling and evaluation of results. Interferences as well as beryllium history such as acid usage, fires, or other high heat situations, recycling materials and methods that might have introduced beryllium or interferences into the facility should all be documented. The following should be identified and included as part of the site history:[9–10, 12,15,17,27,28,32,39,40]

- **Possible interferences** on beryllium sampling and analysis that may exist or have existed at the site
- **Timeline and locations of beryllium work** that has taken place
- **Timeline and locations of work with interferences**
- **Equipment** used in the work situations described above, where it has been used or located elsewhere, and where it is now located
- **Listing and locations of materials** that are present or have been used by location where the materials are known or suspected to contain beryllium. See Table 3.3 for examples.

3.5 Sampling Safety

Sampling must be performed safely for the protection of the samplers, the work site and its occupants, the integrity of the sampling efforts, and the samples themselves.[1–3,11,16,18,19] To accomplish all of this takes careful and deliberate planning and communication in pre-planning, daily planning, oversight, and daily close-out. Thorough communications are more essential on larger complex sampling projects compared with less complex smaller projects.

The sampling project, large or small, must be clearly defined and understood as to what will be done, how, when, where, with what, and by whom. Resource staff such as samplers, record keepers, custodians and shippers of samples, services, and suppliers must be determined and identified.

3.5.1 Personal Protection Equipment

Personal protective equipment (PPE) needs to be carefully and appropriately selected to meet requirements of the requisite code, sampling environments, sampling tasks, and unique characteristics of the individual sampler(s). Ensure that respirators are properly sized, fitted and equipped with protective cartridges, such as HEPA or combination HEPA/other cartridge. Ensure that arms and legs are covered to prevent skin exposures.[1–3,5,18,19] Gloves must be of good quality and correctly sized to the sampler's hand to provide the dexterity required and to avoid fatigue.[16]

3.5.2 Personal Factors and Needs

Personal factors and needs are considerations often overlooked when selecting sampling personnel. Surface sampling requires considerable agility, climbing of

Table 3.3 Materials containing beryllium and materials that are interferences.[a]

Natural materials containing beryllium[14,29]	**Soils**[3]
	Micas
	Hornblende
	Feldspars
	Chlorites
	Vermiculites
	Kandites/kaolinite
	Talc (pyrophyllite)
	Refractory clay
	Woods:
	Alder
	Walnut
	Red Cedar
	Western Red Cedar
	Douglas Fir
	Oregon White Oak
	Yellow Pine
	Sawdust mixtures
	Gemstones: beryl and chrysoberyl gemstones (including aquamarine, emerald and alexandrite)
Man-made materials containing beryllium[14,29,32]	**Manufacturing:** injection molds for plastics, and bearings
	Items created for aerospace: braking systems, bushings, bearings, electronics
	Automotive items: air bag triggers, antilock brake system terminals, steering wheel springs, racing
	Biomedical: dental prostheses, medical laser and scanning electron microscope components, and X-ray windows
	Defense equipment: heat shields, mast-mounted sights, missile guidance systems, nuclear weapon components, tank mirrors
	Energy and electrical: microelectronics, microwave devices, oil field drilling and exploring
	Fire prevention equipment: non-sparking tools and sprinkler system springs
	Instruments, equipment, and objects: bellows, camera shutters, clock and watch gears and springs, commercial speaker domes, computer disk drives, musical instrument valve springs, pen clips, and commercial phonograph styluses
	Sporting goods and jewelry items: golf clubs, fishing rods, and man-made emerald and other gemstones with distinctive colors
	Scrap recovery and recycling: various beryllium-containing products
	Telecommunications: cellular telephone components, electromagnetic shields, electronic and electrical connectors, personal computer components, rotary telephone springs and connectors, and undersea repeater housings
	Cosmetics (consult MSDS for eye shadow and blushes)
	Building materials: brick, refractories ,dry wall, acoustic ceiling tiles

Table 3.3 (*Continued*)

Interferences materials:	Mineral oil
	Common household cleansers
	Chromium
	Iron
	Molybdenum
	Titanium
	Vanadium
	Zirconium

[a]Analysis interferences are method dependent.

ladders, working in cramped congested areas, reaching awkwardly while on ladders at heights, working in hot, humid, noisy and other adverse environmental conditions that directly affect how well sampling personnel actually take a sample. All of these factors must be addressed in protecting the sampler.[16,18,19]

It is important that sampling personnel, and those who perform supervisory and/or oversight duties, understand and fulfill the personal needs of the samplers, such as hydration and rest, in order to prevent and manage unnecessary fatigue. Excessive fatigue can result in sampling without thinking about doing it correctly, which can produce non-defensible and inaccurate samples or worse, an injury or illness.[16] Addressing these quality and safety considerations are even more critical when the sampler has a medical condition such as heart condition, diabetes, epilepsy, vertigo, claustrophobia, simple illness, hangover, or another condition that can affect his or her ability to sample correctly. It is important to know the sampler's condition and ability to sample each day of sampling. Plan of the day team meetings are useful in this regard on campaigns.

Techniques will be discussed later, but one rule of thumb to go by is to perform sampling whenever possible by doing so with one's shoulders above the sample. This will lessen the opportunity for fatigue or muscle twitch setting in while taking a sample.

3.5.3 Sample Protection

Protection of the sample is especially important in the field when the sample is taken while working on a ladder, in a congested space, *etc.* where it is awkward or difficult for the sampler to take the sample and secure it without contaminating it. Working in teams is generally required to collect the sample by the sampler, and securing and doing the paper work by a recorder. The second person also assists in safety activities such as stabilizing the ladder. This increases the assurances of good and reliable samples being obtained safely.[12]

3.6 Recordkeeping

Recordkeeping must be done on every phase of the sampling campaign, so that the data provided by the sampling can be thoroughly evaluated during and after the sampling campaign when archival cross-cutting and other comparisons are performed. All aspects of sampling must be identified in the sampling plan:[1–5,8,15]

- Location of the sampling points
- Individual samples
- Modifications to and deviations from the plan
- Photo log
- Sampling oversight
- Problems with equipment, supplies, laboratory, *etc.*
- Chain of custody
- People problems and any anomalies, special accommodations, *etc.*
- Any other item that might impact on the sampling results.

3.6.1 Chain-of-custody

Chain-of-custody must be documented. Processing of the samples will involve several steps from the point of designation of sample location, taking the sample, securing and protecting the sample, packaging it for sending to the laboratory for analysis, sending it to the laboratory, receiving it at the laboratory, and being analyzed by the laboratory technician.[12]

People handling the samples at each step need to document their involvement through completion of a chain-of-custody form whenever the sample is transferred from one person to another. This serves as proof that the integrity of the sample has not been compromised, ensuring defensibility that the sample has not been tampered with.

3.6.2 Oversight of Sampling

Oversight of sampling (*i.e.* the method, assignments, reports, and forms) is important and should be done, especially on large sampling campaigns. The plan should specify how, or at least that it will be done. Oversight of sampling can be accomplished by many methods, with the key being to document what is observed of the sampling. This information is extremely valuable in assisting and guiding evaluations of the data and decision-making. The more detail that exists about a given sample, the easier it will be to determine whether or not the particular sample or group of samples has acceptable and defendable results. Photographs are especially useful in this regard.[15] Remember, mistakes can be expected to be made by even experienced samplers. Beginners will make more mistakes as they develop their proficiencies. It is important that these are realized, documented, and corrected as needed. Otherwise, the individual sample(s) can produce incorrect results or even place the validity all samples and data of the campaign in question.[15]

3.6.3 Photography Requirements and Permits

Digital photography has made it possible to thoroughly document the sampling process at virtually no cost. A picture is worth a thousand words, and in a sampling campaign, photos can make or break the entire project. The photographer may need some training on what to photograph and how to do it. Ensure that:

- The camera is in good condition
- The photographer knows how to use the camera's settings for close-ups, and how to avoid glare of flash/sun/back lights
- Spare batteries are available with the camera
  Photos should be taken of:
- The outside of the building
- Each overall area being sampled
- Individual close-ups of each sample
- Reference point for the individual sample
- Sampling equipment used such as vacuum cleaners and sampling pumps
- Related sampling activity – storage place of samples awaiting shipment to the laboratory, packaging of the samples, *etc.*
- The individuals who participate in the sampling campaign

Include sample identity information, such as sample number written on the template, in the photograph.

In all cases, the use of photography should be coordinated and approved in advance by the facility organization so that the person taking the photos knows what can and cannot be photographed.

Photos taken related to samples should be referenced on the sampling recordkeeping and oversight forms for future retrieval and evaluation usage. Program the camera to show date and time taken for each photo, and document this information and the name of the photographer on the recordkeeping for each sample.[15]

3.7 Selecting and Pre-qualifying the Laboratory

Selecting and pre-qualifying the analytical laboratory is a must. It is important that the persons selected to perform sampling be detail orientated and possess the "right stuff" to do it consistently and correctly. It is equally important that the selected analyzing laboratory has the "right stuff". First, the laboratory should be accredited by a recognized independent body such as the American Industrial Hygiene Association (AIHA) or demonstrate equivalent proficiency.[1,2] This should be your starting point, and not the end point, in selection of the laboratory. Do not just send samples and accept the result as accurate. Pre-qualify the potential laboratory by sending spiked samples or reference material samples containing beryllium or interferences, of which you know the result, to see if they find it and report accurately.

Require that the laboratory prove that they can, and will, correctly analyze your samples, and report the results to you in a way that provides the degree of accuracy that you require. Some labs may have the ability, but may not do so unless you work with them and demand it of them. Determine with the laboratory what they will need in way of a sample from you in order for them to analyze correctly. Some labs may not have the ability to produce the results you need. For example, what is the smallest amount of a bulk sample needed by the laboratory? Find this out before sampling and sending an actual campaign sample.[45] Sending occasional spiked samples, of which you know the results, before and throughout the sampling campaign as a test of the laboratory will provide you with confidence in the laboratory's analysis.

3.7.1 Quality Control Measures

A key to pre-qualifying and selection of a laboratory is its method of ensuring accuracy and quality of results. Most laboratories have a Quality Assurance Program Plan (QAPP) or quality manual (by whatever name); the latter is a requirement of ISO 17025[47] and will be found in accredited laboratories. Ask for a copy of it, and evaluate closely the laboratory's quality control measures that are followed when analyzing your samples.

Labs report results to customers in different formats with increased details on a graded approach of reports. Determine in advance how much detail you will need on a report, and in what format, when negotiating with the laboratory to analyze your samples before actually selecting the laboratory. The most detailed information, of course, will appear on their best reports. Require that the laboratory demonstrate its performance using reference materials, blanks and low level standards in each analysis batch. Request that the laboratory provides you with raw data, including spectra output, in electronic form so that you can validate that interference corrections were, in fact, done. Do not assume that they will do it correctly; be certain of it with documentation that you can evaluate. This information should be detailed in the sampling plan.

3.8 Sampling Supplies

3.8.1 Consumable Supplies

Sampling involves one-time usage of many items or consumables from personal protection equipment (PPE), sampling area templates, masking tape, strapping tape, markers, hi-liters, computer paper and ink cartridge, bags for waste products, sampling media (camera batteries, brushes, spatulas, pre-moistened wipes such as Ghost Wipes™, pre-weighed micro-pump cassette filters, pre-weighed HEPA filter vacuum bags, 50 mL sample transport vessels with caps, labels, bags, *etc.*).

Table 3.4   Examples of consumables involved with sampling.

- Disposable powder-free gloves (nitrile are recommended)
- Disposable body suits (suit, booties, hood)
- Respirator cartridges (HEPA for beryllium – may need combination for other hazards – discuss with host organization)
- Sample transport vessels (such as 50 ml vials and bags)
- Pre-packaged wetted wipes (such as Ghost Wipes®)
- Wetting agent (solvent, alcohol) for oily surface wipes
- Dry wipes (such as Whatman® filter paper)
- Micro-vacuum pre-loaded cassettes
- HEPA mini-vacuum sampling bags and accessories
- Tape (masking, duct, package sealing strapping, office)
- Labels (write-on and computer generated)
- Pre-printed forms
- Binders (three-ring, other)
- Writing materials (paper tablets, pens, pencils, erasers, white-out)
- Clips (paper clips and spring type clamps)
- Shipping boxes for sending samples to lab
- Permanent marking pens (fine writing and wide tip)
- Templates (100 cm^2 and 512 cm^2 transparency mounts)
- Brushes (for obtaining bulk sweep samples – ensure brush has digestible bristles)
- Disposable PPE (ear plugs; paper suits, booties, hoods, diapers – depending on nature of campaign)
- Bottled water for samplers
- Tacky mats for placement at entrances to restricted areas
- CAUTION or KEEP OUT barricade tape and communication signs to restrict the area being sampled while sampling
- Personal hygiene (soap, showering items – towels, flip flop showering sandals)
- Non-rechargeable batteries (as back-up for recharged types)
- First aid kit and supplies

Necessary consumables need to be identified in the plan with assignments made as to who is responsible for obtaining sufficient supplies, and when these supplies are to be on hand and at what quantity. It is prudent that the inventory status is known at all times, so that adequate supplies are ordered sufficiently in advance to arrive before they are needed. A daily inventory log sheet that identifies what is available, what has been consumed and what has been ordered is advisable. See Table 3.4 for a list of consumable supplies.

3.8.2   Non-consumable Supplies

Non-consumable equipment, and other supplies involved in sampling that will be re-used, also needs to be identified in the plan and secured at the project before it is needed. Some of this equipment will be furnished by contracted samplers, while others may be loaned by the contracting host to contain costs. If you are the loaner, be aware that the equipment could become contaminated through the sampling process. Decontamination costs could make it

Table 3.5 Examples of non-consumables involved with sampling.

Hand tools and other personal tools
- Transportation to and from the work site – private cars or car pool
- Transportation around the site
- Communications (cell phones, radios)
- Scrapers, spatulas
- Laser beam measuring devices
- Tape measures (25 feet with 1" wide steel tape works best)
- Flexible cloth tape measure (for measuring curved surfaces)
- Calculator
- Scissors
- Razor blade box cutters
- Printer capability and supply of paper and inks
- Digital camera with flash cards
- Rechargeable batteries and charger for camera and other equipment)
- Flashlights (hand held and headlamp style)
- Personal protective equipment (hard hat, protective toe footwear, *etc.*)
- Temperature extreme clothing (cool suit/vest)
- Knee pads or kneeling pad
- Respirator and supply of cartridges (HEPA for beryllium – may need combination for other hazards – discuss with host organization)
- Prescription eye protection (can be critical if person cannot see well without or if wearing respirator; full face respirator requires adaptor kit)
- Non-prescription eye protection – goggles, spectacle wraparound, full face shield (face shield works well when sampling above your head)
- Readily available high reach equipment (step ladders, extension ladders, lifts)
- Mini-vacuum cleaner with attachments for bulk sampling

Power equipment, cords and work carts (for each sampling team)
- Supplies from above list depending on sampling locations
- Work cart with at least 3 inch diameter locking wheels for rough surfaces
- Laptop computer and thumb drive portable storage devices
- 50 feet, 110 volt extension cord equipped with multiple plug-in outlet
- Portable GFCI device

economically feasible to buy new or simply designate the item to be used in subsequent remediation efforts. See Table 3.5 for a list of non-consumable equipment.

3.9 Summary

Surface sampling is complex and can be difficult to achieve under the many variables of when, how, what to use, and who is to do what and when. One must carefully determine the reasons for sampling and the expected or desired goals that the sampling is hoped to achieve. Pre-planning can then be done that supports these goals. The result of careful pre-planning can be a plan with prescriptive procedures, so that sampling is done correctly in support of the goals. Thorough planning, recordkeeping and following the plan and procedures ensure that surface sampling is reflective of sound science with legally defensible results.

Acknowledgements

The author thanks the authors of other chapters, and Jim Robbins and Hector Rodriguez of the National Energy Technology Laboratory, Albany, OR, for their assistance and review of this chapter.

References

1. US Code of Federal Regulations, 10 CFR Part 850, *Fed. Regist.*, 1999, **64**, 68854–68914.
2. *Implementation Guide for Use with 10 CFR Part 850*, Chronic Beryllium Disease Prevention Program, US Department of Energy, 2001, DOE G 440.1-7A, www.hss.energy.gov, accessed 16 December 2008.
3. US Code of Federal Regulations, 10 CFR Part 851, *Fed. Regist.*, 2006, **71**, 6858–6948.
4. *Technical Standard, Beryllium-Associated Worker Registry Data Collection and Management Guidance*, US Department of Energy, 2007, DOE-STD-1187-2007, www.hss.energy.gov, accessed 16 December 2008.
5. *Beryllium Fact Sheet*, Sheet Metal Occupational Health Institute Trust, www.smohit.org/pdfs/beryllium_fact_pack.pdf, accessed 19 March 2009.
6. G. E. Whitney, *BH&SC Surface Sampling Tutorial: Basic Methods and Applications*, Los Alamos National Laboratory Document Number LA-UR-07-1868, presented at Beryllium Health and Safety Committee Meeting, March 2007.
7. S. K. Dufay and M. M. Archuleta, *J. Environ. Monit.*, 2006, **8**, 630–633.
8. G. E. Whitney, *Legacy Beryllium in a Multiuse Building*, Los Alamos National Laboratory Document Number LA-UR-04-8368, 2004.
9. D. Field, *Beryllium Legacy Issues at the Nevada Test S*ite, presented at Beryllium Health and Safety Committee Meeting, October 2004.
10. *Beryllium Awareness*, US Department of Energy, 2006, Office of Health Safety and Security Safety Bulletin No. 2006-07, www.hss.energy.gov, accessed 16 December 2008.
11. E. Hewitt, *Employee Protection During Activities in Areas with no Definitive Beryllium Contamination*, presented at 2nd Symposium on Beryllium Particulates and Their Detection, Salt Lake City, UT, November 2005.
12. Method ID-125G, *Metal and Metalloid Particulates in Workplace Atmospheres (ICP Analysis)*. Occupational Safety and Health Administration, 1988 (revised 2002), www.osha.gov/dts/sltc/methods/inorganic/id125g/id125g.html, accessed 11 February 2009.
13. ASTM D 6966-08, *Standard Practice for Collection of Settled Dust Samples Using Wipe Sampling Methods for Subsequent Determination of Metals*, ASTM International, West Conshohocken, PA, 2008.
14. ASTM D7296-06, *Standard Practice for Collection of Settled Dust Samples Using Dry Wipe Sampling Methods for Subsequent Determination of Beryllium and Compoun*ds, ASTM International, West Conshohocken, PA, 2006.

15. G. Rondeau, *CSP Surface Sampling Basics*, presented at 2006 Southeastern Regional Meeting of the American Chemical Society, Augusta, GA, 2006.
16. *TLVs® and BEIs®*, American Council of Governmental Industrial Hygienists, Cincinnati, OH, updated annually.
17. G. A. Day, A. Dufresne, A. B. Stefaniak, C. R. Schuler, M. L. Stanton, W. E. Miller, M. S. Kent, D. C. Deubner, K. Kreiss and M. D. Hoover, *Ann. Occup. Hyg.*, 2007, **51**, 67–80.
18. US Code of Federal Regulations, 29 CFR Part 1910, Subpart Z, Toxic and Hazardous Substances, US Occupational Health and Safety Administration, Washington, DC.
19. US Code of Federal Regulations, 29 CFR Part 1926, US Occupational Health and Safety Administration, Washington, DC.
20. J. P. Cronin, A. Agrawal, L. Adams, J. Tonazzi, M. Brisson, K. White, D. Marlow and K. Ashley, *J. Environ. Monit.*, 2008, **10**, 955–960.
21. ASTM D5438, *Standard Practice for Collection of Floor Dust for Chemical Analysis*, ASTM International, West Conshohocken, PA, 2005.
22. ASTM D 7144-05a, *Standard Practice for Collection of Surface Dust by Micro-vacuum Sampling for Subsequent Metals Determination*, ASTM International, West Conshohocken, PA, 2005.
23. Midwest Filtration Company® Omega™ Vacuum HEPA filters product literature, www.midwestfiltration.com/hepaVacuums.html, accessed 11 February 2009.
24. K. Ashley, G. Applegate, T. Wise, J. Fernback and M. Goldcamp, *J. Occup. Environ. Hyg.*, 2007, **4**, 215–223.
25. K. Ashley, *Standardization Issues in Beryllium Sampling and Analysis: An Update*, presented at 2nd Symposium on Beryllium Particulates and Their Detection, Salt Lake City, UT, November 2005.
26. K. L. Creek, G. Whitney and K. Ashley, *J. Environ. Monit.*, 2006, **8**, 612–618.
27. Notice of Draft Document Available for Public Comment. NIOSH Alert: Preventing Chronic Beryllium Disease and Beryllium Sensitization, *Fed. Regist.*, 2008, 73, 12179.
28. J. Robbins, *Practical Application of PNNL's Visual Sampling Plan to Beryllium Sampling Design*, presented at the Beryllium Health and Safety Committee (BHSC) Meeting, Oak Ridge, TN, March 2007.
29. *Guidance for the Identification and Control of Safety and Health Hazards in Metal Scrap Recycling*, Occupational Safety and Health Administration, Washington, DC, 2008, OSHA 3348-05.
30. Virtual Sampling Plan, Pacific Northwest National Laboratory, Richland, WA. http://dqo.pnl.gov/index.htm. http://vsp.pnl.gov, accessed 19 March 2009.
31. C. B. Davis and N. E. Grams, *When Laboratories should not Censor Analytical Data, and Why*, presented at US EPA's 25th Annual Conference on Managing Environmental Quality Systems, 27 April 2006, Austin, TX.
32. J. L. Jenkins, Jr, *Declaring Victory on Legacy Issues*, presented at the Beryllium Health and Safety Committee Meeting, Las Vegas, NV, 19 October 2004.

33. ASTM D7458, *Standard Test Method for Determination of Beryllium in Soil, Rock, Sediment, and Fly Ash Using Ammonium Bifluoride Extraction and Fluorescence Detection*, ASTM International, West Conshohocken, PA, 2008.

34. A. Agrawal, J. Cronin, A. Agrawal, J. Tonazzi, K. Ashley, M. Brisson, B. Duran, G. Whitney, A. Burrell, T. McCleskey, J. Robbins and K. White, *Env. Sci. Technol.*, 2008, **42**, 2066–2071.

35. NEXTTEQ MICROTEQ® Beryllium ChemTest™ manufacturer's product literature, www.nextteq.com, accessed 11 February 2009.

36. J. S. Duffy, Colorimetric Wipes are Specific, Sensitive, and Fast, *Ind. Hyg. News*, Aug/Sept 2004, www.industrialhygienenews.com, accessed 11 February 2009.

37. R. Song, P. C. Schlecht and K. Ashley, *J. Haz. Mater.*, 2001, **83**, 29–39.

38. K. Ashley, R. Song and P. C. Schlecht, Performance criteria and characteristics of field screening test methods, Am. Lab, June 2002, 32–39.

39. W. G. Cochran, *Sampling Techniques©*, John Wiley & Sons. Inc., Hoboken, NJ, 3rd edn, 1977.

40. J. C. Laul and R. Norman, *J. Chem. Health Safety*, 2008, **15**, 13–25.

41. ASTM E1216, *Standard Practice for Sampling for Particulate Contamination by Tape Lift*, ASTM International, West Conshohocken, PA, 2006.

42. US EPA Statistical software product ProUCL 4.0 for environmental applications for data sets with and without nondetect observations, www.epa.gov/nerlesd1/tsc/images/ProUCL4-Factsheet.pdf, accessed 19 March 2009.

43. L. Maier and M. Van Dyke, Beryllium Education Session, Albany Research Center, presented at US DOE Albany Research Center, 21 June 2005.

44. *Multi-agency Radiation Survey and Site Investigation Manual (MARSSIM)*, US Department of Energy, 2000, www.epa.gov/radiation/marssim/index.html, accessed 24 January 2007.

45. R. A. Brounstein, *Prof. Safety*, 2007 (March), 35–42.

46. P. Wambach and E. Frome, *Statistical Analysis of Beryllium in Settled Dust*, presented at EFCOG/DOE Chemical Management Workshop, 14–16 March 2006, and The R Foundation for Statistical Computing (www.rproject.org).

47. ISO/IEC 17025:2005, *General Requirements for the Competence of Testing and Calibration Laboratories*, International Organization for Standardization, Geneva, 2005.

CHAPTER 4

Sample Dissolution Reagents for Beryllium[*‡]

Applications in Occupational and Environmental Hygiene

KEVIN ASHLEY[a] AND THOMAS J. OATTS[b]

[a] US Department of Health and Human Services, Centers for Disease Control and Prevention, National Institute for Occupational Safety and Health, 4676 Columbia Parkway, M.S. R-7, Cincinnati, OH 45226, USA; [b] BWXT Y-12 National Security Complex, Analytical Chemistry Organization, PO Box 2009, Oak Ridge, TN 37831, USA

Abstract

A variety of sample preparation methods and reagents for beryllium have been used for the dissolution of this element prior to its analytical determination. This chapter provides an overview of methods for beryllium dissolution by

[*] This article was prepared by US Government federal and contractor employees as part of their official duties. The US Government retains a nonexclusive, paid-up, irrevocable license to publish or reproduce this work, or allow others to do so for US Government purposes.

[‡] *Disclaimer*: Mention of company names or products does not constitute endorsement by the Centers for Disease Control and Prevention, the National Institute for Occupational Safety and Health, or the US Department of Energy. The findings and conclusions in this paper are those of the authors and do not necessarily represent the views ot the Centers for Disease Control and Prevention, the National Institute for Occupational Safety and Health, or the US Department of Energy.

Beryllium: Environmental Analysis and Monitoring
Edited by Michael J. Brisson and Amy A. Ekechukwu
© Royal Society of Chemistry 2009
Published by the Royal Society of Chemistry, www.rsc.org

digestion and extraction techniques for a number of sample media, with emphasis on reagents used to prepare samples of interest in the environmental and occupational hygiene fields. Methods normally target the dissolution of the total amount of beryllium originally present in the collected sample. Sample matrices of interest include aerosols, surface samples, ores, soils, beryllium metal, beryllium oxide (including high-fired BeO), and beryllium alloys. Samples of concern for occupational health monitoring include primarily those collected from workplace air and from surfaces. Attributes of the various reagents and techniques are discussed.

4.1   Introduction

In this chapter, sample preparation reagents and dissolution procedures for beryllium are considered, with emphasis on environmental and occupational hygiene samples. There is a wealth of knowledge on sample preparation techniques for subsequent determination of beryllium in a variety of environmental matrices such as airborne particles, soils, sediments, and the like. Unfortunately, much of what is known about sample preparation methods for beryllium in environmental media appears in the older (pre-1980) literature and is often missed by modern-day computer literature searches. In an effort to ameliorate these problems related to the limitations of computerized databases, this chapter strives to provide an overview of some of the important yet "forgotten" literature, while also giving updated information from more recent relevant studies in the subject area. It is not the intention here to exhaustively recount information that has been previously published, but rather to provide a brief survey of the historical literature, along with an update citing more recent investigations in the realm of study.

4.2   Background

In order to measure the beryllium content in samples collected from workplace air, from surfaces in occupational settings and/or other matrices, it is normally necessary to dissolve the collected sample prior to subsequent instrumental measurement of beryllium. Through dissolution, beryllium is ionized and is present in solution as the dication. Techniques to perform this dissolution have historically required acid digestion, many of which have evolved from dissolution procedures for ores, soils, sediments, and other geological sources of beryllium. With these issues in mind, this chapter begins with an overview of sample dissolution procedures for bulk samples, with later discourses on methods for preparing beryllium samples for industrial hygiene monitoring purposes (*e.g.* air, wipe and vacuum samples).

4.3   Beryllium in Geological Media

An overview of the sample preparation methods for geological matrices such as ores and soils is a relevant starting place for discussions of beryllium determinations in environmental media. Dissolution of rock and soil samples is a challenging and often difficult task and, over the years, techniques to decompose and solubilize such media for subsequent determination of beryllium have been investigated and optimized.

4.3.1 Beryllium Ores

Decomposition of beryllium-containing ores such as beryl and bertrandite is ordinarily carried out by means of a fusion process.[1,2] Extremely high-temperature fusion techniques are useful for dissolving refractory materials such as silicates, which are present in high concentration in ores and most other bulk geological media. Sodium carbonate–sodium tetraborate[3–5] and potassium fluoride–sodium pyrophosphate[1] fusions have been shown to quantitatively dissolve beryllium in ores, but treatments with nitric acid alone give low beryllium recoveries from ore samples. Similar fluoride-assisted fusions have been reported for the breakdown of beryl ores and intermediate metallurgical products therefrom.[6–8] Overall, fusion procedures are tedious, time-consuming, and require great skill and extensive experience. Moreover, to complete the dissolution process, it is necessary to carry out subsequent dissolution steps with strong acids such as sulfuric, boric and perchloric acid after the fusion itself is first carried out.

Hydrofluoric acid aids greatly in the dissolution of silicates and is often used in the decomposition of fused ores and other geological specimens. For instance, aqua regia/HF microwave digestion as a first step in the sample preparation process has become routine for the dissolution of beryl.[9] Many of the techniques employing HF that have been found to function well for the breakdown and dissolution of beryllium-containing ores have also been employed for sample preparation purposes on other media of environmental and occupational hygiene interest.

4.3.2 Soils and Silicates

Like ores, soils contain large amounts of silicates (often 50% by weight or greater), and the decomposition of such samples for subsequent beryllium determination requires that these refractory compounds are completely dissolved. This is because the beryllium present is often bound up in the silicate material itself. Hotplate, microwave and similar high-temperature digestions utilizing acid mixtures that include hydrofluoric acid have historically been used for the dissolution of soil samples and related silicate-containing media such as sediments and sludges. High-temperature digestions using mixtures of acids (usually nitric acid plus hydrochloric, perchloric or sulfuric acids, with addition of hydrofluoric acid to facilitate the dissolution of silicates) are commonly used for the dissolution of soils and related samples for subsequent beryllium determination.[10–13]

Because skin exposures in the laboratory, particularly to HF, are extremely dangerous, it is desirable to investigate alternative reagents that may

Table 4.1   Dissolution techniques for beryllium in geological samples.

Matrix(es)	Dissolution Method/Reagents	Reference[a]
Beryl, bertrandite and other ores	(1) $Na_2CO_3/Na_2B_4O_7$ fusion, $>900\,°C$ (2) $HCl/HClO_4$ hotplate digestion	1
Beryl, bertrandite and other ores	(1) $HF/KF/Na_4P_2O_7$ fusion (2) $H_2SO_4/Na_2SO_4/HCl$ digestion	1,4
Ore samples	(1) $Na_2CO_3/Na_2B_4O_7$ fusion, $1000\,°C$ (2) HCl/H_2SO_4 ($+$ $HNO_3/NaOH/$ H_2O_2, if necessary) digestion	3
Beryl and bertrandite ore samples	(1) $Na_2CO_3/Na_2B_4O_7$ fusion (2) HCl/alcohol hotplate digestion	5
Beryl ores	(1) NaF/Na_2CO_3 fusion, $1400\,°C$ (2) NH_4HF_2 treatment ($1000\,°C$)	8
Beryl ores	$HF/HNO_3/HCl/H_3BO_3$ closed-vessel microwave digestion	9
Beryl ores	(1) NaF fusion, $900\,°C$ (2) H_2SO_4/HNO_3 digestion	9
Rock (terrestrial, meteorites and lunar samples)	(1) NaF/Na_2CO_3 fusion (2) HCl dissolution	10
Soils	$HNO_3/HF/HClO_4$ hotplate digestion	11,12
Sediments	HF/H_2SO_4 hotplate digestion	13
Soils and silicate materials	(1) KHF_2/Na_2SO_4 fusion, $\sim800\,°C$ (2) $HClO_4$ (or H_2SO_4) dissolution	14

[a]Details on the various sample preparation procedures can be found in these references.

nonetheless be effective for dissolving silicate materials. Hence, other less hazardous sample preparation methods for soils and related materials have been attempted in which the use of HF was avoided. As an example, an alternative but effective means for decomposition of soil samples entails fusion with a mixture of KHF_2 and Na_2SO_4.[14] This procedure was used as a first step in the separation of ^{10}Be from soil and silicate samples. Additionally, ammonium bifluoride (NH_4HF_2) has been shown to be useful for the dissolution of quartz silicate materials,[15] and is also effective in solubilizing beryllium subsequent to fluoride fusions of bulk samples.[6,8] Also, the utility of NH_4HF_2 for solubilizing beryllium oxide,[16] beryllium hydroxide[16,17] and metallic beryllium[17] has been reported in relevant texts.

Table 4.1 summarizes some of the representative sample preparation procedures used for the dissolution of beryllium in geological samples such as ores and soils. As can be seen from the table, a number of different protocols have been employed and shown to be effective.

4.4   Occupational Hygiene Samples

Of principal concern for the purposes of this overview is the dissolution of beryllium from samples of interest in the occupational hygiene field, with a focus on specimens collected from workplace air and surfaces. In the United States, the most conservative regulatory occupational exposure levels (OELs) for beryllium in air and on surfaces have been established by the US Department of Energy (DOE).[18] Hence for compliance purposes, especially within the DOE complex, it is usually air filters and wipe samples which must be analyzed for beryllium content. This ordinarily requires dissolution of the collected particulate matter plus the sampling medium (*i.e.* the filter or wipe material) using acid digestions. HF is often employed in digestions for dissolving refractory materials such as silicates and beryllium oxide calcined at temperatures in excess of 1500 °C ("high-fired" BeO). More recent work has shown, however, that beryllium extractions with ammonium bifluoride can be successfully performed even from challenging samples, but without dissolving the sampling media (*e.g.* filters or wipes). Methods used to prepare air filters, wipes, and related samples prior to beryllium determination are discussed below.

4.4.1 Workplace Air Samples

Occupational air monitoring of workers potentially exposed to airborne beryllium particles began in earnest in the mid 20th century due to the recognition of this element as a highly toxic workplace hazard.[19] Collection of beryllium aerosols using impingers and air filter samplers have been described.[1,2] Latterly the use of filters for the capture of airborne beryllium, especially those comprised of mixed cellulose ester (MCE), has become the industrial hygiene sampling method of choice for this element.

As to sample preparation methods, early reports described hotplate digestion of cellulosic air filter samples in (primarily) mixtures of nitric and sulfuric acids prior to spectrographic measurement of extracted beryllium.[20,21] Digestion procedures utilizing nitric along with hydrochloric[2] and perchloric[22,23] acids for the dissolution of air filter samples for subsequent beryllium determination were published in later years. A standardized analytical method based on the use of nitric and sulfuric acids (with perchloric acid as an option) for sample preparation was promulgated by the American Industrial Hygiene Association (AIHA) four decades ago.[24] Later, a "tentative" method describing nitric/sulfuric acid digestion for cellulosic filters, and nitric/hydrochloric acid digestion for glass-fiber filters, was published.[25] A related sample preparation method for airborne beryllium collected from emission sources, which relies largely on perchloric acid as a hotplate digestion reagent, was published[26] and field-tested through an interlaboratory trial.[27] Nitric/sulfuric acid digestions are also routinely used for preparation of ambient air and other environmental samples.[28] As hotplate digestion procedures for subsequent beryllium determination, these sample preparation methods are still in wide use to this day.[29]

"Dry ashing" methods, in which quartz-fiber filters,[23] cellulose filters,[29–32] and paper tape[31] samples are ignited at high temperature (500–700 °C) prior to dissolution of the remaining ash in acid, have also been described in earlier papers. These sample preparation procedures, which were originally suitable for colorimetric, fluorimetric or spectrographic analysis, were later applied prior to beryllium measurement by atomic emission spectrometric analysis.[33] They are infrequently used because dry ashing techniques often entail hazardous protocols that can be difficult to carry out and may not be necessary for air sample preparation purposes.

Digestion methods in which HF is employed for dissolution of difficult aerosol matrices, notably refractories such as high-fired BeO, have been reported.[34,35] Because HF is particularly suitable for the dissolution of silicate materials, this acid is used in a French method for dissolving samples collected on quartz-fiber filters.[36]

Leaching of air filter samples with dilute sulfuric acid has been explored for the dissolution of beryllium.[37,38] The procedure is applicable to more soluble beryllium compounds, but is not recommended for insolubles such as refractory beryllium compounds. In a noteworthy application, a sample preparation procedure was proposed in which air filter samples were subjected to boiling in a dilute nitric acid–potassium bisulfate mixture.[39] This method was shown to be effective for dissolving beryllium, including beryllium oxide, in particles over a size range of 1–150 µm.

A sequential extraction procedure for beryllium in air, in which an effort was made to speciate beryllium compounds based on their relative solubilities in chemical reagents of varying strength, has been proposed.[40] The protocol is outlined as follows:

(1) Soluble beryllium salts are dissolved by means of a brief ultrasound treatment in 0.01 M HCl.
(2) Be(II) from metal is then extracted by sonication in 0.1 M copper sulfate.
(3) Be(II) ascribed to BeO is subsequently dissolved by hotplate digestion in a mixture of concentrated nitric and sulfuric acids.
(4) Siliceous Be(II) is dissolved by subjecting the remaining undissolved residue to hotplate digestion in a solution of concentrated nitric and hydrofluoric acids.

Interest in sequential extraction stems from the notion that different metal compounds in air can demonstrate varying toxicities.[41] Further work is required in order to make sequential extraction methods more rugged, and thereby ensure comparability of results obtained from different laboratories.

There are a number of standardized methods for the preparation of airborne beryllium samples collected from workplace atmospheres, and these have been summarized in recent papers.[42,43] An overview of the procedures published by both governmental agencies and consensus standards organizations was provided in these reviews. At the same time, an extensive survey of laboratories performing beryllium-in-air analyses was carried out.[44] Despite the existence of readily available harmonized procedures, it was found that there was little conformity in the reagents which the surveyed laboratories used to prepare air filter samples for subsequent beryllium determination.

Table 4.2 Standardized dissolution techniques for beryllium in workplace air
filter samples.

Method	Dissolution Procedure/Reagents	Reference
NIOSH 7102	HNO_3/H_2SO_4 hotplate digestion	45
NIOSH 7300	$HNO_3/HClO_4$ hotplate or microwave digestion	45
NIOSH 7303	HNO_3/HCl hot block digestion	45
NIOSH 7704	NH_4HF_2 extraction	45
OSHA ID-125G	$HNO_3/HCl/H_2SO_4$ hotplate digestion	46
OSHA ID-206	HNO_3/HCl hotplate digestion	46
HSE MDHS 29/2	HNO_3/H_2SO_4 hotplate digestion	47
INRS Fiche 003	HNO_3/HF ultrasonic extraction	36
ASTM D7035	Options for various acid mixtures; hotplate or microwave digestion	48
ASTM D7202	NH_4HF_2 extraction	51
ASTM D7439	Options for various acid mixtures; hotplate, hot block, or microwave digestion	50
ISO 15202-2	Options for various acid mixtures; hotplate digestion, microwave digestion or ultrasonic extraction	49

Table 4.2 presents a summary of the salient standardized methods used
widely at the present time, along with information about dissolution reagents
recommended, for preparing workplace air samples for subsequent beryllium
measurement.[36,45–51] The methods listed are applicable to occupational air
samples collected using inhalable, thoracic, or respirable samplers.[52] Procedures that utilize nitric and/or hydrochloric acids are applicable to the dissolution of beryllium compounds that are more highly soluble.[53] However,
methods that rely on sulfuric or perchloric acids are required for quantitative
beryllium recovery from refractory matrices such as high-fired beryllium
oxide.[54] Dilute ammonium bifluoride has also been shown to be effective for
dissolving beryllium in air filter samples, including samples containing BeO
calcined at high temperatures.[55–57] Dissolution of silicious material in airborne
particulate samples requires the use of HF[36] or ammonium bifluoride.[15]

4.4.2 Surface Samples

Beryllium in surface dust is usually collected using wipe sampling techniques,[42,58] but vacuum sampling methods are important for dust collection from
non-smooth substrates.[42,59] Cellulosic or polyvinyl alcohol wipe materials are
recommended for sampling of beryllium in surface dust, and a consensus
standard wipe sampling method has been published.[60] For collection of surface
dust from rough or porous surfaces, a consensus standard micro-vacuum
sampling technique has been promulgated.[61] Compared with air filter samples,
samples of surface dust ordinarily contain much more collected material, as

Table 4.3 Standardized dissolution techniques for beryllium in surface samples.[a]

Method	Dissolution Procedure/Reagents	Reference
NIOSH 9102	$HNO_3/HClO_4$ hotplate or microwave digestion	45
NIOSH 9110	NH_4HF_2 extraction	45
OSHA ID-125G	$HNO_3/HCl/H_2SO_4$ hotplate digestion	46
OSHA ID-206	HNO_3/HCl hotplate digestion	46
ASTM D7202	NH_4HF_2 extraction	51

[a]The methods listed are normally applicable to wipe samples, but can also be modified for vacuum and bulk samples collected from surfaces.

well as greater masses of sampling media. Treatment of these materials for sample preparation purposes generally requires greater volumes of acid solution compared with airborne particulate material collected onto air filters.

In a report from over four decades ago,[2] settled dust samples were first treated with nitric acid to destroy any organic material present; sample decomposition was then completed using a potassium fluoride–sodium pyrosulfate fusion. An alternative technique described in an early paper[30] called for (1) dry ashing at 800 °C to destroy carbonaceous material, followed by (2) fusion at 600 °C using potassium bisulfate to ensure complete dissolution of calcined beryllium oxide.

At present, most industrial hygiene laboratories avoid the use of fusion and dry ashing methods for preparing surface dust samples for beryllium analysis, and instead rely mainly on strong acid digestion techniques.[43,44,53] Alternative procedures for preparation of wipe samples, such as boiling in a mixture of dilute nitric acid and potassium bisulfate,[39] or extraction in dilute aqueous ammonium bifluoride,[55–57] have also been proposed. The latter methodology has been standardized and published in the form of an international voluntary consensus standard[51] and also as an approved National Institute for Occupational Safety and Health (NIOSH) analytical method.[45]

A list of standardized methods for preparing samples of surface dust for subsequent determination of beryllium is presented in Table 4.3.[45,46,51]

4.4.3   Bulk Samples

Frequently there is a need to prepare bulk samples for subsequent determination of beryllium. Apart from airborne and surface dust samples, the importance of analyzing soils and accumulated settled dust for beryllium content has arisen at several DOE sites and other locations where nearby soils may have been contaminated with beryllium and/or beryllium compounds. On a number of the DOE sites, there are buildings which may have been used for beryllium work in the past and have been vacant for decades. Such locations are known within DOE as "legacy" sites. In many of the buildings on legacy sites, thick layers of dust with unknown levels of beryllium contamination have built up over the years. To investigate the possibility of environmental sources of beryllium, there is interest in testing samples of the dust from these buildings. These accumulated dust samples can be collected by bulk sampling methods and then analyzed.

The decomposition methods for bulk environmental samples are generally similar to those described above for geological samples such as soil and rock (Table 4.1). In such cases, it is ordinarily necessary to employ a hotplate or microwave digestion technique which uses HF in the acid mix.[62–65] Of course, the aforementioned fusion methods for ore and soil samples may also be applicable. However, fusions must be carried out carefully, for (among other considerations) it is possible that losses of volatile beryllium compounds may occur, for example in samples containing organic material.[66] "Direct" analysis of some matrices can be carried out by spectrometric methods,[67] although a dry ashing procedure may be necessary prior to analysis.[68] In other work, a novel autoclave decomposition technique using xenon tetrafluoride, prior to a sulfuric acid extraction, reportedly gave good recoveries of beryllium from coal and ash samples.[69]

Techniques for decomposing bulk samples such as metals, alloys, and ceramics must be sufficiently robust so as to completely dissolve all of the beryllium present. Again, digestion procedures employing mixtures of nitric and hydrofluoric acids have been demonstrated to be effective for dissolving such challenging sample matrices.[70,71] Nevertheless, alternative dissolution methods have been reported, including the use of concentrated phosphoric acid,[72] ammonium bifluoride,[17,56,57,73] or fluoride fusions.[7,17]

4.5 Summary

The determination of beryllium in samples of interest in the occupational and environmental hygiene arena ordinarily requires that this element is put into solution prior to its analytical measurement. Alternative "direct analysis" techniques have appeared but, generally speaking, dissolution of the sample matrix is necessary to ensure maximal performance with minimal overall analytical uncertainty. A wealth of information on decomposition techniques for subsequent determination of beryllium was published in the latter half of the 20th century, and the knowledge generated has enabled the promulgation of standardized methodologies. Also, newer extraction techniques that are less hazardous than traditional decomposition methods have been shown to be effective for dissolving refractory beryllium samples.

Acknowledgements

We thank Dr M. Hoover for providing an extensive bibliography of historical literature on beryllium. We appreciate the contributions to this effort of the Sampling and Analysis Subcommittee of the Beryllium Health and Safety Committee.

References

1. R. G. Keenan, *Analytic Determination of Beryllium, in Beryllium: Its Industrial Hygiene Aspects,* ed. H. E. Stokinger, Academic Press, New York, 1966, pp. 133–165.
2. R. G. Keenan and J. L. Holtz, *Am. Ind. Hyg. Assoc. J.*, 1964, **25**, 254–263.
3. M. H. Fletcher, C. E. White and M. S. Sheftel, *Ind. & Eng. Chem., Anal. Ed.*, 1946, **18**, 179–183.
4. C. W. Sill, *Anal. Chem.*, 1961, **33**, 1684–1686.
5. R. May and F. S. Grimaldi, *Anal. Chem.*, 1961, **33**, 1251–1253.
6. P. C. Kempchinsky, Beryllium, in *Encyclopedia of Industrial Chemical Analysis,* ed. F. D. Snell and C. L. Hilton, Wiley-Interscience, New York, 1968, pp. 103–141.
7. H. H. Hausner, *Beryllium – Its Metallurgy and Properties,* University of California Press, Berkeley, CA, 1965.
8. S. J. Morana and G. F. Simons, *J. Metals*, 1962, **14**, 571–574.
9. D. S. R. Murty, B. Gomathy and G. Chakrapani, *At. Spectrosc.*, 2000, **21**, 123–127.
10. K. J. Eisentraut, D. Griest and R. E. Sievers, *Anal. Chem.*, 1971, **43**, 2003–2007.
11. W. F. Schmidt and F. Dietl, *Fresenius Z. Anal. Chem.*, 1987, **326**, 40–42.
12. S. Dudka and B. Markert, *Sci. Total Environ.*, 1992, **122**, 279–290.
13. J. R. Merrill, M. Honda and J. R. Arnold, *Anal. Chem.*, 1960, **11**, 1420–1428.
14. J. Stone, *Geochim. Cosmochim. Acta*, 1998, **62**, 555–561.
15. A. R. Timokhin and L. A. Komarova, *Steklo Keram.*, 1985, **6**, 13–15.
16. F. A. Cotton, G. Wilkinson, C. A. Murillo and M. Bochmann, *Advanced Inorganic Chemistry,* Wiley-Interscience, New York, 1999, p. 117.
17. D. A. Everest, *The Chemistry of Beryllium. Elsevier,* Amsterdam, 1964, pp. 3, 38, 39.
18. Chronic Beryllium Disease Prevention Program, *10 CFR Part 850.* US Department of Energy, Washington, DC, 1999.
19. *Recommendations for Control of Beryllium Hazards,* US Atomic Energy Commission, Washington, DC, 1952.
20. J. Cholak and D. M. Hubbard, *Anal. Chem*, 1948, **20**, 73–76.
21. L. E. Owen, J. C. Delaney and C. M. Neff, *Am. Ind. Hyg. Assoc. Q.*, 1951, **12**, 112–114.
22. J. Walkley, *Ind. Hyg. J.*, 1959, **19**, 241–245.
23. D. I. Bokowski, *Am. Ind. Hyg. Assoc. J.*, 1968, **29**, 474–481.
24. Anon, *Am. Ind. Hyg. Assoc. J.*, 1968, **29**, 103–105.
25. R. E. Kupel, M. M. Braverman, J. M. Bryant, A. Carotti, H. M. Donaldson, L. Dubois and E. C. Tabor, *Health Lab. Sci.*, 1971, **8**, 101–106.
26. US Environmental Protection Agency, Method 104. Reference method for determination of beryllium from stationary sources, *Fed. Regist.*, 1973, **38**, 8846–8850.

27. *Collaborative Study of Method 104 – Reference Method for Determination of Beryllium from Stationary Sources*, US Environmental Protection Agency, Washington, DC, 1974, EPA-650/4-74-023.
28. D. D. Thorat, T. N. Mahedevan and D. K. Ghosh, *Ind. J. Chem. Technol.*, 2003, **10**, 67–71.
29. K. Ashley, *Beryllium: Sampling and Analysis,* ASTM International, West Conshohocken, PA, 2006, ASTM STP 1473.
30. R. H. A. Crawley, *Anal. Chim. Acta*, 1960, **22**, 413–420.
31. M. M. Braverman, F. A. Masciello and V. Marsh, *J. Am. Pollut. Contr. Assoc.*, 1961, **11**, 408–409, 427.
32. J. P. McCloskey, *Microchem. J.*, 1967, **12**, 40–45.
33. A. E. Chang, R. Morse, N. H. Harley, M. Lippman and B. S. Cohen, *Am. Ind. Hyg. Assoc. J.*, 1982, **43**, 117–119.
34. R. A. Hiser, H. M. Donaldson and C. W. Schwenzer, *Ind. Hyg. J.*, 1961, **21**, 280–285.
35. A. Zdrojewski, L. Dubois and N. Quickert, *Sci. Total Environ.*, 1976, **6**, 165–173.
36. *Méthodes de Prélèvement et d'Analyse de l'Air*, Institut National de Recherche et de Sécurité, INRS, Paris, 2004, Fiche 003.
37. H. R. Mulwani and R. M. Sathe, *Analyst*, 1977, **102**, 137–139.
38. D. D. Thorat, T. N. Mahadevan and D. K. Ghosh, *Am. Ind. Hyg. Assoc. J.*, 2003, **64**, 522–527.
39. T. J. Miller, *Anal. Lett.*, 1991, **24**, 2075–2081.
40. A. Profumo, G. Spini, L. Cucca and M. Pesavento, *Talanta*, 2002, **57**, 929–934.
41. P. Smichowski, G. Polla and D. Gómez, *Anal. Bioanal. Chem.*, 2005, **381**, 302–316.
42. K. Ashley, M. J. Brisson and S. D. Jahn, *J. ASTM Int.*, 2005, **2**(9), DOI 10.1520/JAI13169.
43. M. J. Brisson, K. Ashley, A. B. Stefaniak, A. A. Ekechukwu and K. L. Creek, *J. Environ. Monit.*, 2006, **8**, 605–611.
44. M. J. Brisson, A. A. Ekechukwu, S. D. Jahn and K. Ashley, *J. ASTM Int.*, 2006, **3**(1); DOI 10.1520/JAI13157.
45. *NIOSH Manual of Analytical Methods*, ed. P. C. Schlecht and P. F. O'Connor, National Institute for Occupational Safety and Health, Cincinnati, OH, 4th edn (with updates), 1994–2006, www.cdc.gov/niosh/nmam/, accessed 11 February 2009.
46. *OSHA Sampling and Analytical Methods*, Occupational Safety and Health Administration Technical Center, Sandy, UT, 2003, www.osha.gov/dts/sltc/methods, accessed 11 February 2009.
47. *Methods for the Determination of Hazardous Substances*, Health and Safety Executive, Sudbury, Suffolk, UK, 1996.
48. ASTM D7035, *Standard Test Method for Determination of Metals and Metalloids in Airborne Particulate Matter by Inductively Coupled Plasma Atomic Emission Spectrometry*, ASTM International, West Conshohocken, PA, 2004.

49. ISO 15202-2, *Workplace Air – Determination of Metals and Metalloids in Airborne Particulate Matter by Inductively Coupled Plasma Atomic Emission Spectrometry – Part 2: Sample Preparation*, International Organization for Standardization, Geneva, 2001.
50. ASTM D7439, *Standard Test Method for Determination of Elements in Airborne Particulate Matter by Inductively Coupled Plasma—Mass Spectrometry*, ASTM International, West Conshohocken, PA, 2008.
51. ASTM D7202, *Standard Test Method for the Determination of Beryllium in the Workplace using Field-based Extraction and Fluorescence Detection*. ASTM International, West Conshohocken, PA, 2006.
52. M. Harper, *J. Environ. Monit.*, 2006, **8**, 598–604.
53. S. Amer, D. Smieja, J. Loughrin and L. Reichmann, L., *J. ASTM Int,*. 2005, **2**(9), DOI 10.1520/JAI13166.
54. A. Stefaniak, M. D. Hoover, G. Day, A. A. Ekechukwu, G. E. Whitney, C. A. Brink and R. C. Scripsick, *J. ASTM Int.*, 2005, **2**(10), DOI 10.1520/JAI13174.
55. E. M. Minogue, D. S. Ehler, A. K. Burrell, T. M. McCleskey and T. P. Taylor, *J. ASTM Int.*, 2005, **2**(9), DOI 10.1520/JAI13168.
56. A. Agrawal, J. Cronin, J. Tonazzi, T. M. McCleskey, D. S. Ehler, E. M. Minogue, G. Whitney, C. Brink, A. K. Burrell, B. Warner, M. J. Goldcamp, P. C. Schlecht, P. Sonthalia and K. Ashley, *J. Environ. Monit.*, 2006, **8**, 619–624.
57. K. Ashley, A. Agrawal, J. Cronin, J. Tonazzi, T. M. McCleskey, A. K. Burrell and D. S. Ehler, *Anal. Chim. Acta*, 2007, **584**, 281–286.
58. S. Dufay and M. Archuleta, *J. Environ. Monit.*, 2006, **8**, 630–633.
59. K. L. Creek, G. Whitney and K. Ashley, *J. Environ. Monit.*, 2006, **8**, 612–618.
60. ASTM D6966, *Standard Practice for Collection of Settled Dust Samples using Wipe Sampling Methods for Subsequent Determination of Metals, ASTM International*, West Conshohocken, PA, 2003.
61. ASTM D7144, *Standard Practice for Collection of Surface Dust by Microvacuum Sampling for Subsequent Determination of Metals*, ASTM International, West Conshohocken, PA, 2005.
62. R. A. Nadkarni, *Anal. Chem.*, 1980, **52**, 929–935.
63. M. A. B. Pougnet, M. J. Orren and L. Haraldsen, *Int. J. Environ. Anal. Chem.*, 1985, **21**, 213–228.
64. M. Bettinelli and U. Baroni, *Int. J. Environ. Anal. Chem.*, 1990, **43**, 33–40.
65. I. Camins and J. M. Shinn, *Analysis of Beryllium and Depleted Uranium: an Overview of Detection Methods in Aerosols and Soils*, Lawrence Livermore National Laboratory, Livermore, CA, 1989, UCID Report No. 21400.
66. M. S. Black and R. E. Seivers, *Anal. Chem.*, 1973, **45**, 1773–1775.
67. T. Nakamura, H. Oka, H. Morikawa and J. Seto, *Analyst*, 1992, **117**, 131–135.
68. E. S. Gladney, *At. Abs. Newsl.*, 1977, **16**(2), 42–43.
69. V. I. Rigin, *Zh. Anal. Khim*, 1984, **39**, 807–812.
70. J. Janoušek, *At. Abs. Newsl.*, 1977, **16**(2), 49–50.

71. T. S. Rudisill and M. L. Crowder, *Sep. Sci. Technol.*, 2006, **41**, 2013–2029.
72. I. Havezov and B. Tamnev, *Fresenius Z. Anal. Chem.*, 1978, **290**, 299–301.
73. A. Agrawal, J. P. Cronin, A. Agrawal, J. C. L. Tonazzi, L. Adams, K. Ashley, M. J. Brisson, B. Duran, G. Whitney, A. K. Burrell, T. M. McCleskey, J. Robbins and K. T. White, *Environ. Sci. Technol.*, 2008, **42**, 2066–2071.

CHAPTER 5

Heating Sources for Beryllium Sample Preparation*‡

Applications in Occupational and Environmental Hygiene

T. MARK McCLESKEY

Los Alamos National Laboratory, MS:J514, Los Alamos, NM 87545, USA

Abstract

There are several forms of beryllium that may be encountered in sampling including beryllium metal, beryllium oxide, and beryllium fluoride, as well as complex beryllium aluminates and silicates in ores. The dissolution of refractory materials often requires aggressive reagents. In order to enhance the kinetics of the dissolution process further, some type of heating is employed in most methods. Heating can accomplished using a hotplate, a furnace, microwave energy, or sonication. Heating can also be performed in an open vented vessel as in the case of fuming acid, or a pressurized container as in the case of pressurized

*This work was authored by an employee of Los Alamos National Security, LLC, operator of the Los Alamos National Laboratory under Contract No. DE-AC52-06NA25396 with the US Department of Energy. The US Government retains a nonexclusive, paid-up, irrevocable license to publish or reproduce this work, or allow others to do so for US Government purposes.
‡ *Disclaimer*: Mention of company names or products does not constitute endorsement by the Department of Energy or its contractors. The findings and conclusions in this paper are those of the authors and do not necessarily represent the views of the Department of Energy.

Beryllium: Environmental Analysis and Monitoring
Edited by Michael J. Brisson and Amy A. Ekechukwu
© Royal Society of Chemistry 2009
Published by the Royal Society of Chemistry, www.rsc.org

microwave digestions. In this chapter, we discuss the different heating techniques that have been employed for soils and industrial hygiene samples.

5.1 Introduction

The previous chapter discussed the variety of decomposition reagents used in various methods for the detection of beryllium. In this chapter, we discuss the different options for heating used to enhance the dissolution rates of beryllium. Several different forms of beryllium are encountered in sampling including beryllium metal, beryllium oxide, beryllium fluoride, and the more complex beryllium silicates found in ores. The high-fired beryllium oxide used in beryllium ceramics and the natural beryllium silicates are especially challenging to dissolve. In order to enhance dissolution rates and ensure that all of the beryllium in a sample is dissolved, all of the methods described previously involve heating as part of the dissolution process. Heating can be accomplished through use of a hotplate, a furnace, a commercial microwave oven, and sonication to enhance beryllium dissolution. Other factors that influence dissolution during heating include the final temperature, the use of an open system *vs.* a closed pressurized system, and the time the system remains at elevated temperature. This chapter focuses on the heating approaches that have been used to detect beryllium in soil and occupational hygiene samples.

Environmental and industrial hygiene samples have different demands in terms of the matrix and particle sizes that are typically encountered. Soil samples have a complex matrix, often including background beryllium in the form of the highly refractory beryllium silicates. When analyzing such systems for beryllium contamination from beryllium processing or beryllium activities, a wide range of beryllium forms and particulate sizes can be encountered. Furthermore, the purpose of the sampling and analysis must be considered; for example, the intent to quantify the baseline beryllium in the soil or to quantify only the added beryllium from anthropogenic activities. In contrast, occupational hygiene samples from air monitoring have few matrix issues outside of the filter paper or cassette used for collection, and the focus is on a well-defined set of particle sizes in the range of 0 to 10 microns. The form of beryllium may be known based on the work activity being sampled. Surface swipes are of intermediate difficulty in that the filter matrix is limited, but if the swipe is collected from an outdoor or dirty environment, many of the same beryllium forms as in soils and a wide range of particle sizes may be present. The mass of material on the swipe is far less than in a soil sample, which makes the dissolution easier.

5.2 Background

In order to obtain accurate information on soil contamination levels and industrial hygiene samples, it is important to ensure dissolution of beryllium in all of its forms. The dissolution and detection of beryllium have been studied for many years. Dissolution rates are inherently sensitive to the species of beryllium at the surface interface, the surface area of beryllium exposed to solution, and mass transfer effects. The beryllium speciation is an important factor.

Beryllium metal dissolves readily in dilute acid, with the generation of hydrogen from solution.[1] Beryllium oxide and beryllium silicates are some of the more difficult species to dissolve, and require more aggressive reagents and reaction conditions. Most sampling and analysis requires the ability to dissolve these relatively inert species. The thermodynamics of dissolution of beryllium oxide have been studied in depth using theory-based models.[2] Dissolution requires two protons per beryllium atom, and experimental studies as a function of pH confirm the second order nature of the dissolution with respect to the proton concentration per unit area of the oxide surface.[3] Decreasing the pH from 5.7 to 2.3 has been observed to increase the dissolution rate by a factor of five.[3] When the pH is lowered further, the proton concentration per unit area of the oxide surface hits a limiting value as the surface becomes saturated with protons.

The dependence of the dissolution rate on protons has lead to reagents that primarily focus on concentrated acids as described in Chapter 4. Most of the dissolution reagents from Tables 4.1, 4.2 and 4.3 include the concentrated acids – HCl, HNO$_3$, HClO$_4$, H$_2$SO$_4$, HF and mixtures thereof. Once the surface of the oxide has a maximum concentration of protons, further enhancement of the dissolution rate must introduce new mechanisms or alter mass transfer processes. The dissolution mechanism may be changed by the introduction of ligands. Ligand assisted dissolution can significantly enhance dissolution rates by changing the thermodynamics of the process through the formation of stable metal ligand complexes in solution. For this reason, the anion of the acid used in the dissolution can play an important role as it complexes to the Be^{2+} in solution. In the presence of both protons and ligands, dissolution rates are maximized as the protons assist in removing O^{2-} from the surface of the BeO lattice to form water and the ligand binds the Be^{2+} to form a stable soluble species.

Ligand assisted dissolution has been demonstrated at moderate pH with a variety of ligands.[4,5] In the case of beryllium, ligand enhanced dissolution under acidic conditions is most pronounced with F$^-$ as the ligand. Most other acids lead to the formation of Be(H$_2$O)$_4$$^{2+}$. Fluorine is the right size to penetrate oxygen defects in the BeO lattice, and the binding constants for the formation of BeF$^+$ and BeF$_2$ are high enough that these species are prevalent even in acidic conditions of pH <2.[6] This effect is made use of in the dissolution procedures that involve HF or ammonium bifluoride.

Further enhancements in the dissolution rate are possible by taking advantage of the Arrhenius behavior for a reaction rate constant and increasing the temperature. Since the dissolution of beryllium material will lead to an increase in entropy as beryllium ions go into solution, the dissolution is helped both by the lowering of the free energy according to $\Delta G = \Delta H - T\Delta S$ (where: ΔG, ΔH and ΔS correspond to the overall free energy, enthalpy and entropy, respectively, for the dissolution process; and T is temperature) and the increase in the rate of dissolution according the Arrhenius equation $k = Ae^{-Ea/RT}$ (where: k = the rate constant; A = pre-exponential factor; Ea = activation barrier; R = gas constant; and T = temperature) as the temperature is increased.[7] Obviously, higher temperatures can only be achieved through some type of heating during the sample preparation/dissolution process. For beryllium dissolution, the addition of heat in the form of traditional conductive heating with a hot plate or furnace, a commercial microwave oven, and sonication have all been examined and are found in use in different published methods. Heating is important to ensure complete dissolution of beryllium materials, and to increase dissolution rates to minimize sample preparation time and allow rapid sample analysis.

5.3   Beryllium in Geological Media and Soils

Beryllium soil samples represent some of the most challenging materials to dissolve. It has been shown that the beryllium minerals often found as background in many soils are especially slow to dissolve. The beryllium silicate $BeSiO_4$, commonly known as phenakite, has a dissolution rate at pH 2 of 2.5×10^{-15} mol cm^{-2} s^{-1} that is over two orders of magnitude slower than the dissolution rate of $MgSiO_4$ (7.5×10^{-13} mol cm^{-2} s^{-1}) and six orders of magnitude slower than $CaSiO_4$ (2.0×10^{-9} mol cm^{-2} s^{-1}).[8] These dissolution rates result in very slow weathering rates for natural beryllium materials. The backgrounds of naturally occurring beryllium have been reported to range from 0.1 to 40 µg g^{-1} soil samples,[9] with typical values in the 0.5 to 3.5 µg g^{-1} range.[10]

In order to obtain accurate data on the total beryllium concentration in a soil sample, it is necessary to both dissolve the refractory beryllium minerals, and break up the soil matrix to some extent to allow access to interior beryllium surfaces. Table 4.1 indicates the heating extremes that have been used to dissolve geological samples. Most of the techniques for dissolution of beryllium in geological samples involve fusions at temperatures ranging from 800 °C to 1400 °C.[11–14] The fusion processes are generally used when one wants to extract all of the beryllium from the soil to get accurate background levels.

In addition to these high temperature fusions for beryllium in geological samples, there are several US Environmental Protection Agency (EPA) analytical methods that list beryllium as an analyte, including Methods 200.7, 3015, 3050B, 3051, 6010, 7090, and 7091.[9] These methods apply primarily to beryllium or beryllium containing solids, or dissolved beryllium in drinking water, waste water, or surface water.

EPA Methods 3050B and 3051 are both specific for solid samples and can be applied to sediments, sludges, and soils.[15] These methods are not fusions, and are typically used to determine contaminant levels in soils. The digestion can be followed by a variety of analytical techniques including flame atomic absorption, graphite furnace atomic absorption, inductively coupled plasma atomic emission spectroscopy, and inductively coupled plasma mass spectrometry. The major distinction between the two methods is in the heating process. Method

3050B allows for microwave heating as well as a hot block or hotplate. The heating involves an acid digestion at 95 °C for 15 minutes, 30 minutes of refluxing with concentrated nitric acid, and a third heating to 95 °C for two hours. Method 3051 is a microwave digestion method. The heating involves raising the temperature to 170–180 °C for 15 minutes in a microwave pressure vessel with a pressure relief valve at 110 psi.

The differences in the procedures for the two methods highlight the options that are possible with microwave heating. Microwaves are readily absorbed by water and the microwave energy is rapidly converted into heat. In the closed vessel microwave system, microwaves can rapidly reach high temperatures and shorten the digestion time. The time saving is a definite advantage. There are also other aspects that must be considered such as potential venting of a high pressure system possibly containing beryllium, and the potential of cross-contamination since the microwave pressures vessels must be re-used. All of these considerations factor into the selection of a method for beryllium analysis.

Most methods that analyze for beryllium are general techniques that have been developed for a suite of metals and there is little accessible published material on detecting beryllium specifically, especially in regards to speciation. For example, most analytical spikes and test samples use beryllium from solution as opposed to the refractory high-fired beryllium oxide encountered in some processing environments. One of the reasons for this is the significant challenge of spiking soils with small amounts of beryllium oxide in a homogeneous manner. There is one recent report on a fluorescent method for the measurement of trace beryllium in soils[16] and a subsequent publication on the inter-laboratory evaluation of the technique using beryllium oxide spiked soils. This technique uses 3% ammonium bifluoride solution as opposed to concentrated acid, and therefore requires a much different heating profile. Heating is performed in a furnace or hot block at 90 °C for 40 hours.[16] At these times, beryllium is extracted from the soil matrix itself as demonstrated by effective recoveries of with US National Institute of Standards and Technology (NIST) Standard Reference Material (SRM) soils containing background beryllium. The same method has been shown to be effective in dissolving BeO spiked soils, with 100% recovery ±15% and a relative standard deviation (RSD) <0.15 in an intra-laboratory study and RSD values <0.1 for a inter-laboratory study using five different spiked levels of beryllium spanning a range from 2.4 to 246 µg Be per gram of soil.[17] The dissolution reagents in this study are milder than either the fuming nitric acid and 30% peroxide of the EPA Method 3050B or the concentrated nitric acid at elevated pressure and temperature used in EPA Method 3051, but, as a result, the heating time at 90 °C with 3% ammonium bifluoride is significantly longer. Both EPA methods and the ammonium bifluoride dissolution either allow or require the filtration of the solution prior to final analyses; this is based on the assumption that the soil matrix will not always be completely dissolved.

There are no large studies comparing the hotplate and microwave digestion techniques used in methods such as EPA Methods 3050B and 3051 for the specific dissolution of beryllium; however, there is a report on comparing these types of dissolution techniques for the detection of lead.[18] An exhaustive study by the environmental lead proficiency analytical testing program examined results from 14 rounds of testing with four samples each involving over 300 laboratories. They compared results from the US National Institute for Occupational Safety and Health (NIOSH) 7082/7105 method using a nitric acid/hydrogen peroxide hot plate dissolution, similar to EPA Method 3050B, to those obtained from the EPA Method 3051 microwave digestion method. They found no large biases when comparing results from hotplate digestion to the microwave digestion.[18]

5.4 Occupational Hygiene Samples

In comparison to the soil samples discussed above, occupational hygiene samples have fewer matrix issues. Workforce air samples are typically collected in areas where airborne beryllium is a concern. In general, such areas are clean from a housekeeping perspective and may even be HEPA filtered. As a result, the air pulled through the filter or cassette for sampling often does not have a challenging background matrix and the filter in almost all cases is a mixed-cellulose ester matrix.

Surface wipes, collected for housekeeping and contamination control, lie in between workforce air samples and soil samples in terms of the complexity of the matrix. The material used to collect surface wipes is more varied than in the case of air sampling, with different methods allowing for different types of wipes including various types of cellulose and Ghost Wipes™. The amount of material on the swipe can vary dramatically depending on the condition of the surface being sampled, and compared to an air sample, may have a large amount of particulate matter. Although analysis of both air samples and surface wipes for beryllium must be capable of determining all forms of beryllium speciation to ensure worker safety, the methods are classified into the two separate areas based on the significant differences in sampling procedures (see Tables 4.2 and 4.3). The focus of occupational hygiene levels is to determine if beryllium levels are above or below certain regulatory threshold values that have changed over time. As the regulatory limits for beryllium exposure become lower and lower, and cases of sensitization persist in the workplace, the concern for dissolving all of the beryllium has surfaced as a concern in terms of the potential variability of particulate size and the challenge in dissolving the high-fired beryllium oxide. Good communication between the industrial hygienist and the analytical laboratory as to the nature of the beryllium source and the possible interferences that may be present, especially in the case of surface swipes, will help insure that the appropriate dissolution procedure is used.

5.4.1 Workplace Air Samples

Most of the methods used to determine beryllium in workplace air samples listed in Table 4.2 involve a hotplate digestion. In terms of heating, this means that the temperature is specified by the nature of the acid, but the time duration may be somewhat variable depending on the operator performing the digestion.

NIOSH Method 7102 specifies the temperatures of 150 °C and 400 °C, which correspond to stop points above the boiling points of each of the component acids used in the 1 : 1 nitric acid : sulfuric acid mixture.[19] In cases where the heating temperature is not defined but the dissolution involves heating a concentrated acid to dryness, the temperature can be roughly determined based on the nature of the acid. The boiling points of the most commonly used acids range from concentrated HCl at 110 °C, 50–70% nitric acid at 122 °C, perchloric acid at 203 °C, to sulfuric acid at 338 °C. The mixtures used in the hotplate digestions that are heated to dryness will reach a variety of final temperatures, and changing a procedure by switching from sulfuric acid to nitric acid has the potential for significant consequences as the temperature during the dissolution will be lowered by over 200 °C.

Most methods still involve fuming acid digestions on a hotplate or microwave digestion in a closed vessel. There are a few methods that have attempted to go to lower heating temperatures and thus reduce the hazards associated with fuming acid procedures which must always be performed in an approved, ventilated fume hood. NIOSH Methods 7303 and 7704 for air samples are methods that use temperatures below the boiling point of water.[19] NIOSH Method 7303 uses concentrated nitric acid at a temperature of 95 °C for 30 minutes, and NIOSH Method 7704 uses a 1% ammonium bifluoride solution heated to 80 °C for 30 minutes. The results reported for NIOSH 7704 show quantitative results for mixed cellulose ester (MCE) filters spiked with beryllium sulfate and recoveries of >95% for filters spiked with beryllium oxide.[20,21]

Another less widely used technique to assist dissolution is ultrasonic extraction (also known as sonication). Ultrasonic extraction refers to the use of ultrasound energy that can be applied either with a bath or a probe. Ultrasound energy is deposited onto a surface in the form of cavitation which produces a bubble that grows and then collapses back on itself. The collapsing cavitation bubble generates high temperatures and pressures in a very small local environment, although the bulk temperature of the solution is not raised significantly. In organic chemical systems, these high local temperatures and pressures can lead to the formation of reactive radicals. The collapsing cavitation bubble also has mechanical effects at the surface of particles, as a liquid jet is formed which then implodes on the surface. This imploding liquid jet can decrease the size of large particles, and expose fresh surfaces or break apart aggregations of smaller particles suspended in solution.[22,23] Both the local heating and the mechanical effects on the particle surface can lead to enhanced dissolution rates.

Probes are capable of greater radiation amplitude and can apply more energy to a given system, but the probe must be placed in direct contact with the dissolution solution. This can lead to cross-over issues. Most results on ultrasonic extraction use probes,[24,25] although there is a report on the effective extraction of Cu, Pb and Zn from soil samples using an ultrasound bath.[26]

Dissolution can also be enhanced by simple mechanical rotation or stirring to eliminate mass transfer effects. One recent report compares mechanical agitation, heating to 80 °C and sonication using 1% ammonium bifluoride as the dissolution medium.[20] In this report, mechanical agitation, heating to 80 °C and sonication all lead to quantitative extraction of $BeSO_4$. In the case of BeO, recoveries varied from >90% for heating to >83% for mechanical agitation and sonication. The combination of heating and sonication was not reported. Presumably sonication and heat would combine the advantages of increasing the bulk temperature with the mechanical forces exerted on the particle surface from bubble cavitation and collapse.

5.4.2 Surface Samples

Surface samples span an amazing wide range of potential matrix effects which may even exceed those encountered in soil samples. Swipes may have little more than a few particulates when sampling fresh clean surfaces, or be embedded with significant amounts of soil when sampling outside, or be contaminated with a wide range of man-made oils and machining fluids when sampling historic sites ready for decommissioning. As a result, both the industrial hygienist and the analyst must be aware of potential contaminants for the best analytical results.

Microwaves can be borne in either a closed or open system. The closed system is typically associated with microwave digestions and allows for high temperatures, but can be problematic when carbonates or significant amounts of organics are present that lead to over-pressurization. The potential hazard of a beryllium containing sample over-pressurizing must be addressed for these methods. Microwaves can also be used in an open system as a fast convenient way to heat the sample quickly and to maintain a given temperature, with no danger of the over-pressurization that could be encountered in some surface samples. The open vessel microwave system has the advantages that it allows for effective automation, evaporation to dryness, and when properly vented, enables the use of HF if needed for the more difficult samples that may be encountered with surface wipes.[27]

Table 4.3 lists the standardized methods for the detection of beryllium in surface samples. NIOSH Method 9102 uses a perchloric acid/nitric acid mixture in conjunction with heating at 150 °C.[19] Three of the methods involve hotplate acid digestions in which the exact temperature may not be mentioned but is limited to the fuming point for the specific acid mixture. The other two closely related methods, ASTM D7202 and NIOSH 9110, both involve heating to 80 °C for 30 minutes with 1% ammonium bifluoride.[19,28]

As recently as 2006, NIOSH Method 7303, which was originally presented as valid for elemental beryllium, was extended to be valid for beryllium sulfate and shown to have an 80% recovery for high-fired beryllium oxide.[29] Also in 2006, a theoretical approach for evaluating analytical digestion methods for poorly soluble particulate beryllium was presented.[30] This report brought to light concerns regarding variation in particle size that could affect the dissolution rates for digestion procedures in occupation hygiene sampling for beryllium. Workforce air samples are generally concerned with airborne particles in the range of 0–10 μm that can be readily deposited in the lungs, whereas surface wipes have the potential to encounter a wide range of particle sizes. A follow-on experimental study using bulk quantities of BeO demonstrated the importance of heating with recoveries for BeO of 27% at ambient conditions and 90% when heated to 90 °C.[31] Lower recovery was also reported for larger particle sizes in the one example presented. This type of study emphasizes the need to test methods with challenging samples such as BeO, and the need for reference materials with varied particle sizes in addition to the SRM 1877 with 200 nm BeO particles recently developed by NIST. Methods must also be tested at the relevant spike levels. Bulk studies on dissolution may not accurately reflect the dissolution times or the temperatures needed for accurate analytical sampling, which is focused on very low levels of beryllium materials.

The current US Occupational Safety and Health Administration (OSHA) personal exposure limit of $2 \mu g\, m^{-3}$ and the DOE swipe levels of 0.2 and $3 \mu g/100\, cm^2$ used to release equipment and maintain housekeeping require the capability to spike filters consistently with BeO solid at μg levels to accurately asses the dissolution methods. Ongoing efforts at the US Department of Energy Y-12 site to develop BeO solid spiked filters for analytical testing have resulted in an effective procedure for spiking filters with known quantities of BeO from a suspended solution of BeO prepared using an ultrasonic bath.[16,17] This has enabled more recent beryllium-specific methods to perform inter-laboratory validations with BeO solid as well as the more traditional beryllium from solution. These solid spiked filters have the potential to help determine the required temperatures and times for heating in dissolution procedures especially when, as often occurs, dissolution procedures are modified to some degree from the original method.

Fundamental dissolution studies could also be performed on thin films of BeO. This has been done in the case of silica dissolution. Silicon and beryllium both have a strong affinity for fluoride and react with ammonium bifluoride in a similar manner. Silicon reacts as SiO_2 according the equation: $SiO_2 + 4NH_4HF \rightarrow (NH_4)_2SiF_6 + 2NH_4F + 2H_2O$, in the same manner that BeO reacts: $BeO + 2NH_4HF \rightarrow (NH_4)_2BF_4 + H_2O$. The reaction of SiO_2 has been studied as a function of temperature by monitoring the change in thickness of a glass plate exposed to the ammonium bifluoride solution.[32] This study shows that a change in temperature from 20 °C to 80 °C using a 5% ammonium bifluoride solution results in an increase in the dissolution rate by a factor of 12.6 with a rate of dissolution of $22.5 \mu m\, h^{-1}$. The flat surface in some respect represents the worst case scenario of a very large particle being dissolved at the surface.

5.5 Summary

There are four types of heating used in current beryllium detection methods. Fusion at very high temperatures is used to determine background beryllium levels in soils. Detection methods to determine contamination levels can use hotplate acid digestion, microwave digestions or sonication extraction.

Sonication extraction does not heat the bulk system, but can produce high effective temperatures in the local environment. Microwaves provide rapid bulk heating of an aqueous system and can be used in an open system to obtain the same effective temperature as a hot plate or furnace. Microwaves may also be used with closed pressure systems to obtain a combination of high temperature and moderate pressure to enhance dissolution. The largest comparison of hotplate digestion and microwave digestion has been performed using digestion methods similar to EPA-Method 3050B and Method 3051; this found no statistical difference between the two dissolution methods for the recovery of lead. Most hot plate methods involve fuming acid digestions. NIOSH Methods 7303 and 7704 for air samples use temperatures below the boiling point of water to avoid fuming conditions. If refractory materials such as BeO are anticipated, then it is important to know that the heating temperatures and times in the digestion method have been shown to have effective recoveries with BeO spikes.

References

1. F. A. Cotton and G. Wilkinson, *Advanced Inorganic Chemistry,* Wiley-Interscience, New York, 1980, p. 276.
2. M. A. Gomez, L. R. Pratt, J. D. Kress and D. Asthagiri, *Surf. Sci.*, 2007, **601**, 1608–1614.
3. G. Furrer and W. Stumm, *Geochim. Cosmochim. Acta*, 1986, **50**, 1847–1860.
4. E. Bauer, D. Ehler, H. Diyabalanage, N. N. Sauer and T. M. McCleskey, *Inorg. Chim. Acta*, 2008, **361**, 3075–3078.
5. C. Ludwig, W. H. Casey and P. A. Rock, *Nature (London)*, 1995, **375**(6526), 44–47.
6. L. Alderighi, P. Gans, S. Midollini and A. Vacca, *Adv. Inorg. Chem.*, 2000, **50**, 109–172.
7. J. H. Espenson, *Chemical Kinetics and Reaction Mechanisms,* McGraw Hill, New York, 1981, p. 156.
8. W. H. Casey and H. R. Westrich, *Nature*, 1992, **355**(6356), 157–159.
9. T. P. Taylor, M. Ding, D. S. Ehler, T. M. Foreman, J. P. Kaszuba and N. N. Sauer, *J. Environ. Sci. Health, Part A*, 2003, **38**, 439–469.
10. T. Asami and M. Kubota, *Environ. Geochem. Health*, 1995, **17**, 32–38.
11. M. H. Fletcher, C. E. White and M. S. Sheftel, *Ind. Eng. Chem., Anal. Ed.*, 1946, **18**, 179–183.
12. S. J. Morana and G. F. Simons, *J. Metals*, 1962, **14**, 571–574.
13. D. S. R. Murty, B. Gomathy and G. Chakrapani, *At. Spectrosc.*, 2000, **21**, 123–127.
14. J. Stone, *Geochim. Cosmochim. Acta*, 1998, **62**, 555–561.
15. SW-846, Test methods for Evaluating solid waste, Revision 6 Feb 2007, 3rd edn, 2007, http://www.epa.gov/epawaste/hazard/testmethods/sw846/pdfs/toc.pdf
16. A. Agrawal, J. P. Cronin, A. Agrawal, J. C. L. Tonazzi, L. Adams, K. Ashley, M. J. Brisson, B. Duran, G. Whitney, A. K. Burrell, T. M.

McCleskey, J. Robbins and K. T. White, *Environ. Sci. Technol.*, 2008, **42**, 2066–2071.

17. J. P. Cronin, A. Agrawal, L. Adams, J. C. L. Tonazzi, M. J. Brisson, K. T. White, D. Marlow and K. Ashley, *J. Environ. Monit.*, 2008, **10**, 955–960.

18. P. C. Schlecht, R. G. Song, J. H. Groff, H. A. Feng and C. A. Esche, *Am. Ind. Hyg. Assoc. J*, 1997, **58**, 779–786.

19. *NIOSH Manual of Analytical Methods*, ed. P. C. Schlecht and P. F. O'Connor, National Institute for Occupational Safety and Health, Cincinnati, OH, 4th edn (with updates), 1994–2006, www.cdc.gov/niosh, accessed 11 February 2009.

20. A. Agrawal, J. Cronin, J. Tonazzi, T. M. McCleskey, D. S. Ehler, E. M. Minogue, G. Whitney, C. Brink, A. K. Burrell, B. Warner, M. J. Goldcamp, P. C. Schlecht, P. Sonthalia and K. Ashley, *J. Environ. Monit.*, 2006, **8**, 619–624.

21. K. Ashley, A. Agrawal, J. Cronin, J. Tonazzi, T. M. McCleskey, A. K. Burrell and D. S. Ehler, *Anal. Chim. Acta*, 2007, **584**, 281–286.

22. F. Priego-Capote and M. D. L. de Castro, *J. Biochem. Bioph. Methods*, 2007, **70**, 299–310.

23. A. J. Saterlay and R. G. Compton, *Fresenius' J. Anal. Chem.*, 2000, **367**, 308–313.

24. E. C. Lima, F. Barbosa, F. J. Krug, M. M. Silva and M. G. R. Vale, *J. Anal. At. Spectrom.*, 2000, **15**, 995–1000.

25. A. V. Filgueiras, J. L. Capelo, I. Lavilla and C. Bendicho, *Talanta*, 2000, **53**, 433–441.

26. R. Al-Merey, M. S. Al-Masri and R. Bozou, *Anal. Chim. Acta*, 2002, **452**, 143–148.

27. V. F. Taylor, A. Toms and H. P. Longerich, *Anal. Bioanal. Chem.*, 2002, **372**, 360–365.

28. ASTM D7202, *Standard Test Method for the Determination of Beryllium in the Workplace using Field-based Extraction and Fluorescence Detection*, ASTM International, West Conshohocken, PA, 2006.

29. S. Amer, D. Smieja, J. Loughrin and L. Reichmann, in *Beryllium: Sampling and Analysis*, ASTM International, West Conshohocken, PA, 2006, pp. 62–67, ASTM STP 1473.

30. A. B. Stefaniak, C. A. Brink, R. M. Dickerson, G. A. Day, M. J. Brisson, M. D. Hoover and R. C. Scripsick, *Anal. Bioanal. Chem.*, 2007, **387**, 2411–2417.

31. A. B. Stefaniak, G. C. Turk, R. M. Dickerson and M. D. Hoover, *Anal. Bioanal. Chem.*, 2008, **391**, 2071–2077.

32. A. R. Timokhin and L. A. Komarova, *Glass Ceram.*, 1985, **42**, 267–269.

CHAPTER 6

Beryllium Analysis by Inductively Coupled Plasma Atomic Emission Spectrometry and Inductively Coupled Plasma Mass Spectrometry*‡

Applications in Occupational and Environmental Hygiene

MELECITA M. ARCHULETA[a] AND BRANDY DURAN[b]

[a] Sandia National Laboratories, P.O. Box 5800, MS, 0871, Albuquerque, NM 87185-0871, USA; [b] Los Alamos National Laboratory, PO Box 1666 MS:J514, Los Alamos, NM 87545, USA

*This article was prepared by US Government contractor employees as part of their official duties. The US Government retains a nonexclusive, paid-up, irrevocable license to publish or reproduce this work, or allow others to do so for US Government purposes.
[a] Sandia National Laboratories is a multiprogram laboratory operated by Sandia Corporation, a Lockheed Martin company, for the United States Department of Energy's National Nuclear Security Administration under contract DE-AC04-94AL85000.
[b] Los Alamos National Laboratory is operated by Los Alamos National Security, LLC, under Contract No. DE-AC52-06NA25396 with the US Department of Energy.
‡ *Disclaimer*: Mention of company names or products does not constitute endorsement by the Department of Energy or its contractors. The findings and conclusions in this paper are those of the authors and do not necessarily represent the views of the Department of Energy.

Beryllium: Environmental Analysis and Monitoring
Edited by Michael J. Brisson and Amy A. Ekechukwu
© Royal Society of Chemistry 2009
Published by the Royal Society of Chemistry, www.rsc.org

113

Abstract

A variety of analysis methods have been developed for the identification and quantitative analysis of beryllium and beryllium compounds. The current state-of-the-art for most metal analysis and, in particular beryllium analysis, is inductively coupled plasma atomic emission spectrometry (ICP-AES) [also known as inductively coupled plasma optical emission spectroscopy (ICP-OES)] and inductively coupled plasma mass spectrometry (ICP-MS). This chapter provides an overview of ICP-AES and ICP-MS as they relate to the identification and quantitative analysis of metals and in particular beryllium. Issues discussed include: instrumentation available; interferences such as spectral, matrix, and physical interferents; and specific considerations when working with beryllium.

6.1 Introduction

The fundamental principle used in inductively coupled plasma (ICP) technologies is the use of a high temperature plasma discharge to generate excited atoms and positively charged ions. The sample, typically in liquid form, is converted to an aerosol and transported into the base of the plasma torch, where it travels through the different heating zones of the plasma. Here it is dried, vaporized, atomized, ionized, and transformed into a fine liquid aerosol that arrives at the plasma as excited atoms and ions that represent the elemental composition of the sample. The excitation of the outer electron of a ground-state atom to produce wavelength-specific photons of light is the fundamental basis of atomic emission spectrometry (AES). There is also enough energy in the plasma to remove an electron from its orbital to generate an ion. It is the generation, transportation, and detection of significant numbers of these positively charged ions that give ICP-MS its characteristic ultra trace detection capabilities.

In general, the ICP methods are applicable to elemental analysis of essentially all of the elements in the periodic table. In addition, this technology offers simultaneous or rapid sequential multi-element analytical capability within a broad dynamic range at relatively good precision, without changing the operating conditions. This chapter discusses some of the similarities and differences between ICP-AES and ICP-MS technologies, specifically as they relate to beryllium analysis. Issues such as sample preparation for ICP-AES and ICP-MS, interferences, and quality control are discussed.

6.2 Preparation of Samples

A wide variety of sample preparation methods have been developed for the dissolution of beryllium and beryllium compounds, and are described in Chapter 4. Many of these methods work equally well for the digestion of samples prior to analysis by ICP methods. Normal methods that can be used for ICP work utilize acids to dissolve the sample into solution so that it is available for analysis.

One of the challenges with identifying an appropriate method for preparing samples for beryllium analysis is ensuring that the preparation method is capable of digesting the various forms of beryllium that will be encountered as well as the samples matrix (*i.e.* wipe or filter). Of particular concern is beryllium oxide (BeO), which is readily formed on airborne beryllium particles and on beryllium metal objects. This form of beryllium and low-fired beryllium oxide are relatively easy to digest by most of the methods described in Chapter 4. However, high-fired beryllium oxide, which is manufactured today, is more difficult to digest and analyze.[1] Most published digestion methods were developed using soluble standards of metals and have not necessarily been tested for all possible compounds of the analyte. For this reason, it is crucial that all candidate sample preparation methods be verified with respect to their suitability for dissolving elements of interest from the particular materials which could be present in the test atmosphere.

6.2.1 Methods Available for Sample Analysis by ICP-AES or ICP-MS

If results are required for total beryllium, including beryllium oxide and other insoluble beryllium compounds, it is necessary to select a suitable sample preparation method that has been determined to be robust enough to digest these compounds. Some compounds, such as refractory beryllium oxides, require a more robust sample preparation method than is required for other compounds of beryllium, or for the metals or metalloids themselves. A list of standardized dissolution techniques is provided in Chapter 4 (Table 4.2). If results are required for insoluble beryllium, dissolution methods utilizing H_2SO_4, $HClO_4$, HF, or NH_4HF_2 are appropriate. Alternatively, if it is known that no insoluble compounds of beryllium are used in the workplace, and that none are produced in the processes for which sampling is being carried out, it is possible to employ a method utilizing HCl and/or HNO_3 to prepare solutions for ICP-MS analysis and/or ICP-AES analysis.

6.2.2 Analytical Considerations for Selecting a Sample Preparation Method

When selecting a sample preparation method, it is important to take into consideration the applicability of each method for dissolution of any other metals or metalloids of interest that could provide interferences depending on the digestion protocol and detection method used. If there is any doubt about whether the selected sample preparation method will exhibit the required analytical recovery when in the presence of other compounds or metalloids, it is crucial to determine its effectiveness for the particular application prior to utilizing the method. There are several ways to accomplish this.

For total metals and metalloids, analytical recovery may be estimated by analyzing a performance evaluation material of known composition that is similar in nature to the materials being sampled in the workplace and contains the potential interfering compounds also expected to be present. An example evaluation material would be a representative certified reference material (CRM). For beryllium, several reference materials are available from a number of suppliers. In addition, a standard reference material (SRM) for high-fired beryllium oxide is currently available from the US National Institute of Standards and Technology (NIST), SRM 1877 Beryllium Oxide Powder.[2] These certified reference materials can be used individually and together to establish the effectiveness of preparation methods.

There are several considerations when using bulk samples. A bulk sample has certain physical characteristics, such as particle size and agglomeration, which could have a significant influence on the efficacy of its dissolution. Also when working with powders and bulk samples, smaller quantities of material are often much more easily dissolved than larger quantities. If during sample preparation, undissolved residue remains, further sample treatment may be required in order to dissolve target analyte elements. This could entail filtration to capture the undissolved material, or subsequent digestion of the residue using an alternative sample preparation method.

For soluble metals and metalloids, analytical recovery is best determined by analyzing spiked media blanks (*i.e.* filters spiked with solutions containing known masses of the soluble compound(s) of interest). In order to be confident of your analytical preparation method, recovery should be at least 90% of the known value for all elements included in the spiked media blanks, with a relative standard deviation of $<5\%$.[3] If the analytical recovery is outside the required range of acceptable values, the use of an alternative sample dissolution method may be warranted.

6.2.3 Challenges with Beryllium Samples for Analysis by ICP-AES or ICP-MS

In addition to the issues associated with digesting insoluble compounds of beryllium, the analytical technique used to identify and quantify beryllium has limitations depending on the dissolution method used. There are many published methods for the preparation of metals which include beryllium samples in various matrices. The details are discussed in Chapter 4. Of particular concern are extractions with hydrochloric, hydrofluoric, perchloric, or sulfuric acids. While these acids are found to be effective dissolution acids for beryllium and its compounds, they exhibit problems and issues such as interferences, material compatibility, and personnel safety that need to be addressed if used for digestion of samples to be analyzed by ICP-AES or ICP-MS.

Extraction in mixtures of hydrochloric and nitric acids has been shown to be effective for the dissolution of numerous metals and metalloids, including beryllium, present in air filter samples.[4] In addition, hydrochloric acid is an effective solvent for many metal oxides, phosphates, sulfides, and basic silicates, while nitric acid is an oxidizing agent that effectively dissolves many metals and metalloids and their compounds. This acid mixture, however, is not effective for the dissolution of acidic silicates and some metal oxides, particularly BeO, that are resistant to acid attack. Furthermore, digestion with hydrochloric acid introduces a matrix effect that interferes with the ability to analyze beryllium by ICP-MS (see Section 6.5.1).

Extraction in mixtures of sulfuric acid and hydrogen peroxide has also been shown to be effective for the dissolution of numerous metals and metalloids, including beryllium, present in air filter samples.[4] Sulfuric acid is effective for dissolution purposes, owing in part to its high boiling point (340 °C). This facilitates the decomposition of substances that may not break down at lower temperatures. Some elements (*e.g.* Ba, Ca, Pb) may form insoluble sulfates, which are alleviated by the addition of hydrochloric acid.

Although sulfuric acid and hydrogen peroxide are not effective for the dissolution of silicate materials and some metal oxides that are resistant to acid attack, they have been found to be effective with beryllium oxide. Sulfuric acid however, poses significant safety concerns when using microwave digestion methods. The use of sulfuric acid in a microwave digestion method has been found to cause fires due to the higher temperatures and pressures needed for an effective dissolution. In addition, sulfuric acid matrices can contribute to an increase in background noise, resulting in decreased sensitivity, and changes in the viscosity of the samples resulting in transportation issues to the plasma. When using sulfuric acid, matrix matching of the standards and sample becomes important.

A third extraction method using mixtures of nitric acid and perchloric acid has been shown to be effective for the dissolution of numerous elements, including beryllium, from airborne particulate matter.[4] Perchloric acid is a strong oxidizing agent and solvent, and is especially useful for dissolving ferroalloys. However, neither nitric acid or perchloric acids, nor their mixtures, are effective for the dissolution of silicate materials. The addition of hydrochloric acid can aid in the dissolution of certain elements (such as Te) from especially difficult sample matrices. Perchloric acid fumes can collect and crystallize in the hood's ventilation system, causing explosive conditions if care is not taken to keep the system running efficiently.

The use of hydrofluoric acid (HF) is necessary for the dissolution of metals and metalloids that are bound to silicate materials, and may be required for refractory metal oxides such as beryllium oxide. However, it is often not the acid of choice in the analytical laboratory due to its safety concerns. The major concern working with HF is the potential for exposure. Hydrofluoric acid is corrosive and a contact poison. Because of its low dissociation constant, it easily and quickly penetrates the skin destroying the tissues in the body. Death can occur if as little as 2.5% of total body surface area is exposed to concentrated HF.[5,6] Furthermore, hypocalcaemia and hypomagnesaemia can occur from a smaller surface area or lower concentrations. Additional training and personal protective equipment, along with the presence and training required for the use of calcium gluconate gel, may be required for working with HF in the laboratory. Furthermore, if hydrofluoric acid is employed in sample preparation, it is necessary to use corrosion-resistant laboratory ware made from materials that are not attacked by HF [*e.g.* polytetrafluoroethylene (PTFE)], as well as platinum cones in the ICP-MS. In addition, a different introduction system for ICP-AES is needed, such as a v-groove or high-dissolved solids type nebulizers that are HF resistant. The disadvantage of such types of nebulizers over the standard glass concentric nebulizer is an increase in detection limits.

6.3   Quality Control and Quality Assurance

As part of a quality assurance (QA) program, the laboratory should maintain and establish quality control (QC) acceptance criteria for all methods. Quality control should include the accuracy and precision of all analyses performed and QC samples in duplicate to estimate the uncertainty of measurement of all calibrations and test methods. The availability of a certified reference material is important, not only for the verification of digestion or dissolution methods as described earlier, but also for the confirmation of the quality of both the analysis and the digestion methods during the analysis. As a minimum, there are three types of quality control samples that must be run when analyzing samples by ICP-MS and ICP-AES.

For the calibration of the instrument, a calibration curve should be generated with a minimum of three calibration standards that bracket the expected sample concentrations and a calibration blank.[3] For ICP-AES, the standards should be made up in the same acid matrix as the samples.[7] The calibration curve linearity must initially be verified by preparing and analyzing a check standard. This is an independent standard (from neat materials) or a standard from an independent source at a concentration that is at the high level of the calibration curve. This check standard should analyze within a 5% difference from the actual standard concentration. Instrument calibration should be verified every 24 hours or each time the calibration curve is generated, whichever is less. In some cases, the established methodology sets requirements on continuing calibration verification to verify instrument stability. This could require that the check standards be run as often as every 10 or 20 samples.

Minimum reporting limits for the element of interest should be established by analyzing media spiked samples, prepared at the desired minimum reporting limit concentrations, and taken through the entire analytical process. Instrument performance at the minimum reporting limit concentration is established and verified with each analytical batch through the analysis of another analytical standard prepared at the analyte's minimum reporting limit concentration.

Multiple matrix-based quality control spikes, also independently prepared, should also be analyzed with each batch of samples. The spike level should be at a concentration level within the calibration curve of the applicable analysis. These laboratory control samples are generally carried through the entire procedure, from preparation to analysis, and are run at the beginning and end of the sample batch or as required by the analytical method. The purpose of these quality control samples is to establish and verify the accuracy and precision of the entire method.

In addition to the quality control samples discussed above, ICP-AES requires one additional quality control. For ICP-AES, an appropriate interference check standard should also be analyzed at the beginning and at the end of each analytical run. These check standards will consist of either known interfering analytes at high concentrations, or analytes that are present in the sample at high concentrations. The resultant analysis of this check standard can be compared with the standard blank to monitor for increased or decreased background. Such samples are analyzed applying the same set of standard calibration data. This analytical standard is used to verify an accurate analyte response in the presence of possible interfering materials present in the samples.[3]

Internal standards and their response and recoveries are sometimes monitored and required to be within certain ranges. This information can be used as part of quality control for sample preparation and analysis.

As a final step to ensure the quality of the analysis methods being used, a proficiency testing program of some sort should be used. The purpose of proficiency testing, or round robin testing, is to assist a laboratory's ability to verify its analytical performance by providing samples on a regular basis, evaluating the results, and providing a report on how well the laboratory performed. For beryllium, the American Industrial Hygiene Association (AIHA) initiated a beryllium proficiency testing program (BePAT) in 2004 to assist laboratories with their performance in analyzing beryllium samples.[8]

6.4 ICP Overview

Since its introduction in the early 1970s, the use of inductively coupled plasma discharge for use in emission spectrometry (ICP-AES) and mass spectrometry (ICP-MS) has rapidly expanded worldwide in terms of number of users, applications, manufacturers, and types of instrumentation. Deciding which instrument to use is not always simple. ICP-AES offers rapid analysis for multiple elements in complex matrices with moderate sensitivity, whereas, ICP-MS offers improvements over other techniques by combining the rapid multi-element capability of the ICP-AES with the low detection limits of graphite furnace atomic absorption (GFAA). Furthermore, ICP-AES offers the ability to correct for interfering substances by choosing alternative wavelengths. Most elements have multiple wavelengths available by ICP-AES. On the other hand ICP-MS often only has a few isotopes that can be used for the analysis and in the case of beryllium only one, ^{9}Be.

It is somewhat surprising, however, that the current use of ICP-MS technology has not grown as rapidly as ICP-AES, despite improvements in the area of mass spectrometry. When we compare the history of ICP-MS and ICP-AES, we see that for the same period of time, five times more ICP-AES systems are in use worldwide today than ICP-MS systems.[9] Clearly one of the reasons is price; an ICP-MS system typically costs twice as much as an ICP-AES system and three times as much as a GFAA, although the price is becoming more competitive. Another reason is that the ICP-MS has a reputation of being a complicated research piece of equipment, and thus is generally saved for the research laboratory and not the analytical laboratory. And, third, although ICP-AES has more limitations due to interferences, ICP-MS is much more sensitive to matrix effects. We must understand and respect both systems if we are to meet the new demands and requirements for increased sensitivity for beryllium analysis.

For either system, ICP-AES or ICP-MS, the sample introduction system is very much the same. A sample material in solution is introduced through a nebulizer, where it is converted into a fine aerosol with argon gas. The fine droplets of the aerosol ($< 10 \, \mu m$) are separated from the larger droplets through a spray chamber and transported into the plasma torch where energy transfer processes cause the desolvation, volatilization, and atomization of the sample constituents.[10] In ICP-MS, the positively charged ions are extracted from the plasma through a vacuum interface and separated on the basis of their mass-to-charge ratio and detected. In ICP-AES, the plasma generates photons of light *via* excitation of electrons. AES measures the wavelength-specific photons that are emitted from the sample as the electrons return to ground state.

The fundamental design of an ICP sample introduction system has not changed dramatically since the technique was first introduced in 1983. The main function of the sample introduction system is to generate a fine aerosol of the sample. The mechanism of introducing a liquid sample into analytical plasma can be considered as two separate events: a nebulizer converts the liquid to a fine spray; and a spray chamber sorts the droplets that are to enter the plasma. The sample is generally introduced into the nebulizer *via* a peristaltic pump that pumps the samples at a rate of $\sim 1 \, mL \, min^{-1}$.[10]

Several options are available for sample introduction, each with its own significant source of random and system error in the measurement of samples by ICP. There are several considerations when selecting the proper sample introduction system: dissolved solids content; suspended solids; presence of strong acids (specifically sulfuric and hydrofluoric acid); detection limit requirements; precision requirements; sample load requirements; sample size limitations; and interferences in general.[11]

Because the plasma discharge tends to be inefficient at dissociating large droplets, the spray chamber is designed to sort the droplets so that only the small droplets enter the plasma. It also works to smooth out pulses that occur during the nebulization process, due mainly to the peristaltic pump. There are many different designs of spray chambers available, all working to allow only the smallest of droplets into the plasma for dissociation, atomization, and ultimately ionization.

The plasma source is generated by a magnetic field created from radio-frequency (RF) power applied to the load coils surrounding the torch system.

Argon gas is introduced into this magnetic field where electrons are stripped from the argon atoms, accelerated within the magnetic field, and thus creating an inductive coupling. The fine sample aerosol is then introduced into this plasma. In ICP-AES, the light emitted by the excited atoms is converted to an electrical signal which is then converted to a concentration, whereas in ICP-MS, the ions are separated by their mass to charge ratio which can also be measured quantitatively.[12]

6.5 Analysis by ICP-AES

A variety of ICP-AES systems are commercially available. Variations in systems include: the torch viewing (radial, axial, or dual); simultaneous or sequential instruments, depending on the type of wavelength dispersions; and a large variety of detector types (photodiode array, charge-injection device, or charge coupled device). Each variation comes with its own unique set of advantages and disadvantages. In general, the decision to choose one type of instrument over another is largely influenced by the price and operating costs, the sample output, and the required sensitivity levels. Despite all of these differences, the overall operation of the instrument is the same as are the analytical issues associated with achieving a good analysis.

As discussed, the preparation of beryllium samples is highly variable depending on the starting materials and the selected preparation method. The final digestate matrix of the sample will influence the outcome of the analysis. ICP-AES is unique in that complex matrices can be analyzed by this technique, with the caveat that the calibration standards are matrix matched with the samples to minimize matrix interferences and emission suppression. This is further discussed below.

Instrument parameters may also vary between manufacturers, but the following are typical starting points for method development. Power settings for water matrices are typically defaulted to 1 kW, whereas high salt content requires a higher setting of 1.2 kW, and organic matrices may need an even larger power increase of 1.4 kW. Unless sample volume is minimal, a minimum of three replicates per analyte should be conducted to provide valid statistical data. In addition, a rinse between samples is crucial for minimizing sample carryover, and thus time settings should be verified. Most instruments will have a humidifier for the argon gas used in the nebulization of the sample, which aids in transport of samples containing salts or high dissolved solids. Lastly, an instrument warm-up time, as recommended by the individual manufacturer, should be followed to ensure plasma stability and minimize spectral drift. Usually this ranges from 20 to 30 minutes.

6.5.1 Interferences

Several types of interferences may be encountered during an analysis, which can lead to inaccurate results. Spectral interferences can be caused by background emission noise, stray light from high concentrations elements, or overlap of spectra from nearby analytes.

Influence from background noise can be minimized by matrix matching the calibration standards and by conducting background corrections.

Spectral overlap corrections are a little more complicated. Depending on the instrument type, one could simply choose an alternate wavelength. If an alternate wavelength is not available, then an inter-element correction can be used. Most instruments have the capability to perform this correction automatically, but it is up to the analyst to determine the correction factor to be used, for it will vary between instruments due to the differences in resolution. It will also vary between matrix types and with changes in instrument parameters. A correction factor for a 5% nitric acid matrix is not likely to be the same for a 10% nitric or a dilute nitric and hydrochloric matrix, nor will it be the same if operating the torch at 1200 kW *versus* 1000 kW. Therefore, the analyst must verify this correction if sample or operating conditions change.

The actual correction factor is determined by analyzing a known concentration of the interferent analyte, analyzing a matrix "blank" (analyte-free solution), and then calculating the intensity contribution of the interferent on the selected wavelength. The interferent is then analyzed with the desired analyte, and an interferent subtraction is performed to determine the resultant analyte concentration.

Physical interferences and matrix effects can also lead to inaccurate analysis. These interference effects are mainly associated with the sample introduction system. Changes in viscosity and surface tension can cause significant inaccuracies. Effects from high dissolved solids and acid concentrations can be reduced by sample dilution, a change in the nebulizer type, or by wetting the argon prior to nebulization. Again, matrix matched standards will help to alleviate these interferences.[13]

When low detection limits are needed and matching complex matrices is difficult, an internal standard can be used to correct this problem. For maximum precision and accuracy, internal standards should be used for any and all samples. This process entails the addition of an analyte not present in the sample, which the instrument uses as a means of adjusting intensity fluctuation caused by matrix changes. The key to using this correction is that the analyst verifies that the analyte chosen as an internal standard is not present in the sample and is not interference. Internal standards can be used to correct for slight short term noise and drift.

Another type of physical interference is referred to as a memory effect. This is a result of carryover of an analyte from the previous sample, which then contributes to the signal in a new sample. This can result from an inadequate rinse time and build-up of a sample material on the uptake tubing or spray chamber. The required rinse time can be determined by running sequential blanks following a sample analysis and observing the decrease in background signal. Many new instruments incorporate this monitoring automatically (known as a smart rinse), in which rinsing is continued between samples until the signal returns to baseline levels.

In all cases, interferences can be determined by calibrating a method using water or dilute acid matrix standards, then performing an analysis with the sample matrix blanks and with spiked blanks, and analyzing single element, high concentration standards to determine if the intensity signals change in comparison to the standard blank or if false positive or negative concentrations are detected by the single element standards.

It is important to note that many samples may contain uncommon elements or elements not required for analysis. If these analytes are present at high concentrations in the sample, then they too must be verified as a potential interferent. Though these elements may not display a spectral interference, many element combinations can result in an elemental suppression or expression, and thus should be investigated.

Most instruments have the capability to perform a full spectral scan which allows the analyst to see qualitatively the various analytes in a sample, thus allowing for investigation of possible interferences.

6.5.2 Considerations when Working With Beryllium

Many spectral libraries are available to consult when determining which wavelength to choose during method development. Most instruments already have these libraries built into their operating system and have predefined the preferred wavelengths. For beryllium there are three prominent wavelengths: 234.86 nm; 313.04 nm; and 313.11 nm. Many of the sample preparation methods, such as National Institute for Occupational Safety and Health (NIOSH) Method 7301,[14] recommend a wavelength to use such as 313.04 nm.

One thing to be aware of is that the 313.04 nm and 313.11 nm wavelengths are very close together and will appear in the same spectra window as a doublet. In addition, there is an OH doublet on either side of the beryllium spectrum. If the instrument does not calculate an automatic background correction, then the analyst must be cautious not to set the background correction on these additional peaks which would result in an overcorrection. The estimated reported detection limit for beryllium on the 313.04 nm wavelength is 0.2 ng mL^{-1}.[14] The 313.04 nm wavelength has known spectral interferences from vanadium and titanium. Again, further investigation of other sample elements should be conducted to determine if additional interferences are detected.

The 234.861 nm wavelength is an alternate to the 313.042 nm wavelength. Estimated intensity is approximately 10% less intense than the 313.042 nm line. The 234.861 nm wavelength has reported interferences of iron and titanium. The detection limit for beryllium at the 234.861 nm wavelength is 0.31 ng mL^{-1}.

If the ICP-AES is a simultaneous instrument, then the operator may choose to run all three wavelengths during an analysis. Multiple line analysis may not be an option for operation of sequential instruments, since this entails a longer run time and more sample solution.

Though either line has reported interferences, these are typically determined by analysis of high concentration, single element standards at concentrations greater than 100 parts per million (ppm). Such analytes are not an issue if they are not present in the sample at high concentrations, or the concentration of beryllium is several magnitudes above background where there is minimal contribution from these elements.

Since beryllium concentrations in samples can vary greatly depending on the source of the sample, a dual calibration may be needed to ensure accuracy within a large concentration range. Most instruments will allow for multi-line analysis in which the analyst can set two calibration curves, one for low detection at the parts per billion (ppb) range and a higher calibration curve in the ppm range. A single curve can be used over a large concentration range, but calculation of such curves tend to skew towards the higher standards, and thus if one needs detection at the lower end, such a range may cause a loss in detection limits. By setting up two individual calibration curves, one eliminates this possibility in addition to eliminating the need to dilute higher samples and re-running this analysis. This allows the operator to perform the analysis and choose which reported value to use depending on where it falls within the two calibration curves.

As stated previously, ICP-AES allows for a rapid analysis of multi-elements in complex matrices. For beryllium, detection limits in the low ppb range are achievable. But as described in Chapter 1, the need for lower detectable concentrations may exceed ICP-AES capabilities and thus analysis using mass spectrometry may be necessary.

6.6 Analysis by ICP-MS

The mass spectrometer is generally the ICP detector of choice when sensitivity is an issue. In the mass spectrometer, the plasma is used to generate positively charged ions rather than photons as in ICP-AES. ICP-MS generates such a large number of positively charged ions that detection limits range in the parts per trillion compared with ICP-AES which has detection limits in the parts per billion range.

The heart of the system (the mass separation device – sometimes called the mass analyzer) is the region of the ICP mass spectrometer that separates the ions according to their mass-to-charge ratio (m/z). This selection process can be achieved in a number of different ways depending on the mass separation device. In the classical type of mass spectrometer, a charged particle passes through a magnetic field, and is deflected at an angle proportional to the mass and charge of the particle to separate the ions of interest from all the other non-analyte, matrix, solvent, and argon-based ions.

For the first 10 years after its development, ICP-MS used traditional quadrupole mass filter technology to separate the ions of interest. This worked exceptionally well for most applications, but proved to have limitations in determining difficult elements or dealing with more complex sample matrices. This led to the development of other mass separation devices such as the double-focusing magnetic-sector design. The main limitation of this technology is the slow turn around time due to the scanning and settling of the magnet.

Another newer technology, time-of-flight (TOF) spectrometry, was developed to compensate for the inherently sequential nature of the quadrupole. The main attribute of TOF technology is its unique ability to sample all ions generated in the plasma at exactly the same time, which is advantageous when large amounts of data need to be captured in a short amount of time.

A number of elements have a poor detection capability by ICP-MS because of major spectral interferences generated by ions derived from the plasma gas or the reagents used in the digestion. The collision reaction cell or ion trap was developed to work around these problems. A problem with this technology is that it is sometimes difficult to optimize and is only suitable for a few of the interferences. It is also susceptible to more severe matrix effects. The ion trap, which is very similar to the quadrupole, also has problems with scattering of ions due to the abundance of argon species in the ion beam.

6.6.1   Selectivity and Interferences

In the case of beryllium, where detection limits mandate that selectivity be high, it is important to consider the significance of any known interferences in the context of the measurement task.

Interferences are generally classified into three major groups: spectral; matrix; and physical.

6.6.1.1   Spectral Interferences

Spectral interferences are those interferences where the spectra of two elements are so similar as to be indistinguishable from each other. Two important performance specifications of a mass analyzer govern its ability to separate an analyte peak from a spectral interference. The first is resolving power or resolution; the second is abundance sensitivity. Even though they are somewhat related and both define the quality, abundance sensitivity is probably the most critical.

Abundance sensitivity is the degree to which the signal of a mass peak contributes to adjacent masses. The abundance sensitivity is affected by ion energy and quadrupole operating pressure. Signal interferences may result when a small ion peak is being measured adjacent to a large one. The potential for these interferences when other elements are expected in the sample should be recognized, and the spectrometer resolution adjusted to minimize them. If a mass analyzer has good resolution but poor abundance sensitivity, it will often prohibit the measurement of an ultra trace analyte from a more abundant interferent element (*i.e.* small peak next to a major interfering mass).

Spectral interferences can also come from the matrix or gas used in the sample introduction system, or by entrained oxygen and/or nitrogen from the surrounding air. One example of this is the combination of oxygen and the hydrochloric acid media used to dissolve some compounds of beryllium. $^{16}O^+$ combines with the most abundant chlorine isotope at 35 atomic mass unit (amu) to form $^{35}Cl^{16}O^+$, which interferes with the vanadium (^{51}V) often used as an internal standard for beryllium.[9] Another type of spectral interference is produced by elements in the sample combining with H^+, $^{16}O^+$ or $^{16}OH^+$ $^{10}OH^+$ to from molecular hydrides, oxides or hydroxides, or the formation of doubly charged species formed from easily ionizable elements which show up at one-half of the mass that they would normally be seen. Table 6.1 provides a listing of common molecular ion interferences.[4]

Still another type of spectral interference, called isobaric, is produced by different isotopes of other elements in the sample creating spectral interferences at the same mass as the analyte of interest. Such interferences must be recognized as being unavoidable by the selection of an alternative analytical isotope (*i.e.* beryllium does not have an alternative isotope); corrections must be made to the data. To examine the possibility that the matrix could present interference, one must refer to published information and consider the relationship between the magnitude of interferences and the expected or required detection limit of the element to be determined. If matrix interferences are significant, alternatives such as a different mass-to-charge ratio or use of a collision reaction system (if available) should be considered.[11]

6.6.1.2   Matrix Effects

In addition to spectral interferences, ICP-MS has historically been considered highly sensitive to matrix effects. This is considered one of the major drawbacks to using ICP-MS. One of these effects is often called "sample transport effect" and occurs when the dissolved solids or acid concentration in the sample suppresses the analyte signal. This is often caused by the impact of the sample on droplet formation in the nebulizer or, in the case of organic matrices, by variations in the pumping rate of the solvent with differing viscosities. The second type of matrix suppression occurs when the sample affects the ionization conditions of the plasma discharge and again causes suppression.[4] Due to its low mass, beryllium is particularly sensitive to space-charge effects where low mass elements are scattered out of the ion beam by the heavier mass elements, causing a loss of sensitivity.

6.6.1.3   Physical Interferences

Physical interferences generally refer to the physical processes that involve the transport of the sample into the plasma, the sample conversion in the plasma, and the movement of ions through the plasma–mass spectrometer interface. These interferences generally result in differences between instrument responses for the sample and the calibration standards. Physical interferences may occur as a result of viscosity differences and surface tension (affecting transfer of the solution to the nebulizer, and aerosolization and transport to the plasma), or during excitation and ionization processes within the plasma itself. High levels of dissolved solids and matrices that are not fully digested with the sample may contribute deposits of material on the extraction, or skimmer cones, or both,

Table 6.1 Common molecular ion interferences.

Molecular Ion	Mass	Element Interference[a]
Background Molecular Ions		
NH^+	15	–
OH^+	17	–
OH_2^+	18	–
C_2^+	24	–
CN^+	26	–
CO^+	28	–
N_2^+	28	–
N_2H^+	29	–
NO^+	30	–
NOH^+	31	–
O_2^+	32	–
O_2H^+	33	–
$^{36}ArH^+$	37	–
$^{36}ArH^+$	39	–
$^{40}ArH^+$	41	–
CO_2^+	44	–
CO_2H^+	45	Sc
ArC^+, ArO^+	52	Cr
ArN^+	54	Cr
$ArNH^+$	55	Mn
ArO^+	56	–
$ArOH^+$	57	–
$^{40}Ar^{36}Ar^+$	76	Se
$^{40}Ar^{38}Ar^+$	78	Se
$^{40}Ar_2^+$	80	Se
Matrix Molecular Ions		
Chloride		
$^{35}ClO^+$	51	V
$^{35}ClOH^+$	52	Cr
$^{37}ClO^+$	53	Cr
$^{37}ClOH^+$	54	Cr
$Ar^{35}Cl^+$	75	As
$Ar^{37}Cl^+$	77	Se
Sulfate		
$^{32}SO^+$	48	–
$^{32}SOH^+$	49	–
$^{34}SO^+$	50	V, Cr
$^{34}SOH^+$	51	V
SO_2^+, S_2^+	64	Zn
$Ar^{32}S^+$	72	–
$Ar^{34}S^+$	74	–
Phosphate		
PO^+	47	–
POH^+	48	–
PO_2^+	63	Cu
ArP^+	71	–
Group I, II metals		
$ArNa^+$	63	Cu
ArK^+	79	–
$ArCa^+$	80	–

Table 6.1 (*Continued*)

Molecular Ion	Mass	Element Interference[a]
Matrix oxides[b]		
TiO	62–66	Ni, Cu, Zn
ZrO	106–112	Ag, Cd
MoO	108–116	Cd

[a]Method elements or internal standards affected by molecular ions.
[b]Oxide interferences will normally be very small and will only impact the method elements when present at relatively high concentrations. Some examples of matrix oxides are listed of which the analyst should be aware. It is recommended that Ti and Zr isotopes be monitored if samples are likely to contain high levels of these elements. Mo is monitored as a method analyte.

reducing the effective diameter of the orifices and, therefore, ion transmission. Although dissolved solids levels not exceeding 0.2% (w/v) have proven helpful to reduce such effects, the primary way to compensate for physical interferences is to use an internal standard.

6.6.1.4   Internal Standards

It is important to select an appropriate number and combination of internal standards to correct for instrument drift and physical interferences. For full mass range scans, it is recommended to use a minimum of three internal standards. For the analysis of beryllium, vanadium has been suggested as an internal standard for most beryllium analysis by ICP-AES, but it is generally more important to select an internal standard with a mass ion close to the analyte of beryllium having a similar ionization potential such as lithium.

Internal standards must be selected such that they are suitable for the intended purpose, exhibit adequate sensitivity, are not present in the test solutions, and are chemically compatible with the test solution matrix (*i.e.* they must not cause precipitation).[4] While internal standards may also be used to correct for transport interferences and nebulizer efficiencies that arise from a matrix mismatch between the calibration and test solutions, matching the matrix of the calibration and test solutions is generally preferred.

6.6.1.5   Memory Effects

Memory effects can also be seen in ICP-MS just like in ICP-AES. Memory effects can result from sample deposition on the sampler and skimmer cones, and from the build-up of sample material in the plasma torch and spray chamber. As with ICP-AES, occurrence of memory effects can be minimized by flushing the system with a rinse blank consisting of HNO_3 (1 + 49) (1.000̄) in water between samples. The possibility of memory interferences should be recognized within an analytical run, and suitable rinse times and rinse solutions used to reduce them.

6.6.2 Considerations when Working with Beryllium

The major drawback to using ICP-MS for the analysis of beryllium is its high sensitivity to matrix effects. The mass spectrometer is specifically sensitive to sample transport effects seen with GhostWipe™ samples (Environmental Express, Mt Pleasant, SC). These GhostWipes™ are the common sampling media used for the collection of surface samples for beryllium housekeeping and material transport. The digestion of GhostWipes™ results in a situation where the dissolved solids or acid concentration in the sample suppresses the analyte signal. Furthermore, GhostWipe™ matrix effects can also be physical. This is where the high level of dissolved solids due to the GhostWipe™ matrix is not fully digested with the sample, thus contributing deposits of material on the extraction, or skimmer cones, or both, reducing the effective diameter of the orifices and, therefore, ion transmission. Although the use of multiple extraction methods and filtering can help with these interferent problems, the normal way is to use an internal standard. The use of an internal standard, however, does not completely reduce the physical interferences caused by the sampling matrix, resulting in much more instrument maintenance on a more regular schedule than with ICP-AES.

References

1. M. J. Brisson, K. Ashley, A. B. Stefaniak, A. A. Ekechukwu and K. L. Creek, *J. Environ. Monit.*, 2006, **8**, 605–611.
2. Material Details, SRM 1877 – Beryllium Oxide Powder, National Institute of Standards and Technology, https://www-s.nist.gov/srmors/view_-detail.cfm?srm=1877, accessed 12 February 2009..
3. *prEN 13890, Workplace Atmospheres – Procedures for measuring metals and metalloids in airborne particles – Requirements and test methods*, Comité Européen de Normalization, Brussels, 2007.
4. ASTM D7439, *Standard Test Method for Determination of Elements in Airborne Particulate Matter by Inductively Coupled Plasma-Mass Spectrometry*, ASTM International, West Conshohocken, PA, 2008.
5. E.B. Segal, *Chem. Health Safety*, 2000, **7**, 18–23.
6. L. Muriale, E. Lee, J. Genovese and S. Trend, *Ann. Occup. Hyg.*, 1996, **6**, 705–710.
7. C.B. Boss and K.J. Fredeen, *Concepts, InstrumentationZ and Techniques in Inductively Coupled Plasma Optical Emission Spectrometry*, Perkin Elmer Corporation, USA, 1999.
8. Beryllium Proficiency Analytical Testing (BePAT) Program, American Industrial Hygiene Association, www.aiha.org/Content/LQAP/PT/BePAT.htm, accessed 12 February 2009.
9. R. Thomas, *Practical Guide to ICP-MS. A Tutorial for Beginners*, CRC Press, Boca Raton, FL, 2008.
10. S. Greenfield and A. Montaser, *Inductively Coupled Plasmas in Analytical Atomic Spectrometry*, VCH Publishers, New York, 2nd edn, 1992.

11. H.E. Taylor, *Inductively Coupled Plasma Mass Spectrometry, Practices and Techniques,* Academic Press, New York, 2001.
12. R.K. Winge, V.A. Fassel, V.J. Peterson and M. A. Floyd, *Inductively Coupled Plasma-Atomic Emission Spectroscopy; An Atlas of Spectral Information,* Elsevier, New York, 1985.
13. ASTM D7035, *Standard Test Method for Determination of Metals and Metalloids in Airborne Particulate Matter by Inductively Coupled Plasma Atomic Emission Spectrometry.* ASTM International, West Conshohocken, PA, 2004.
14. Method 7301, in *NIOSH Manual of Analytical Methods,* ed. P. C. Schlecht and P. F. O'Connor, National Institute for Occupational Safety and Health, Cincinnati, OH, 4th edn (with updates), 1994–2006, www.cdc.gov/niosh/nmam/, accessed 12 February 2009.

CHAPTER 7

Beryllium Analysis by Non-Plasma Based Methods*[‡]

ANOOP AGRAWAL[a] AND AMY EKECHUKWU[b]

[a] AJJER LLC, 4541 East Fort Lowell Road, Tucson, AZ 85712, USA; [b] Savannah River National Laboratory, Savannah River Site, Aiken, SC 29808, USA

Abstract

This chapter provides an overview of beryllium analysis methods other than plasma spectrometry (inductively coupled plasma atomic emission spectrometry or mass spectrometry). The basic methods, detection limits and interferences are described. Specific applications from the literature are also presented.

7.1 Introduction

The most common method of analysis for beryllium is inductively coupled plasma atomic emission spectrometry (ICP-AES). This method, along with inductively coupled plasma mass spectrometry (ICP-MS), is discussed in Chapter 6. However, other methods exist and have been used for different

---

*This article was prepared in part by a US Government contractor employee as part of official duties. The US Government retains a nonexclusive, paid-up, irrevocable license to publish or reproduce this work, or allow others to do so for US Government purposes.

[‡] *Disclaimer*: Mention of company names or products does not constitute endorsement by Savannah River Nuclear Solutions (SRNS) or the US Department of Energy (DOE). The findings and conclusions in this paper are those of the author and do not necessarily represent the views of SRNS or DOE.

Beryllium: Environmental Analysis and Monitoring
Edited by Michael J. Brisson and Amy A. Ekechukwu
© Royal Society of Chemistry 2009
Published by the Royal Society of Chemistry, www.rsc.org

applications. These methods include spectroscopic, chromatographic, colorimetric, and electrochemical.

7.2 Fluorescence

7.2.1 Background

Fluorescence spectroscopy (also known as fluorometry or spectrofluorimetry) is a type of electromagnetic spectroscopy which analyzes fluorescence from a sample. It involves using a beam of light, usually ultraviolet light, that excites the electrons in molecules of certain compounds and causes them to emit light of a lower energy, typically, but not necessarily, visible light. Generally, the species being examined will have a ground electronic state (a low energy state) of interest, and an excited electronic state of higher energy. Within each of these electronic states are various vibrational states.

Fluorescence is an ideal method of detection because it is extremely sensitive and non-destructive, and can be performed quickly using inexpensive instrumentation. Further, the fluorescence method is routinely employed in biological industries using high throughput methods. Thus, a fluorescence test to analyze beryllium offers an opportunity to significantly increase the test throughput and reduce cost for high volume users.

Fluorescent detection of beryllium has been reported since the 1950s, with literature reports on a variety of fluorescent indicators including morin,[1–3] chromotropic acid,[4] and Schiff bases.[5] Despite the many reports of fluorescent indicators for beryllium, a complete system for its fluorescence detection was not approved by a regulatory agency. However, recent developments to change the situation targeted three goals simultaneously.[6] These goals were: (a) a dissolution method that is able to dissolve beryllium and beryllium oxide (BeO), and remains compatible with the fluorescence indicator; (b) tolerance to a wide variety of interferences; and (c) a minimal number of simple steps from dissolution to detection. This has resulted in standardized tests approved by both ASTM International[7] and the US National Institute of Occupational Safety and Health (NIOSH) for wipes and air filters,[8] while a test method for soil analysis has also been approved by ASTM International.[9] The discussion below focuses on the development and the efficacy of the standardized tests. Dissolution is described in greater detail in Chapter 4 and a summary is provided below that pertains to the use of dissolution methods adopted for the standardized test procedures.

7.2.2 Applications

7.2.2.1 Dissolution

Typical dissolution methods for the dissolution of BeO from a swipe involve concentrated inorganic acid and heating; in addition, some methods use

Table 7.1 Summary of results for filters spiked with BeO (μg Be/filter).

Expected	Rotated	75 °C	90 °C
0.2	0.19	0.23	0.21
1–1.2	0.65	1.11	1.20
4–5	2.68	4.54	4.20

hydrogen peroxide (see Chapter 4). Such conditions are not compatible with any known fluorescent indicator, so the solution must be evaporated to dryness and further treated before it can be added to the fluorescent indicator. In order to eliminate the time-consuming digestion steps of current standard methods, the use of a fluoride-based medium to dissolve beryllium was investigated. The efficacy of fluoride as a dissolution ligand is described in Chapter 5. It was found that Be metal was dissolved within seconds in 1% ammonium bifluoride (NH_4HF_2; ABF).[6] Results obtained from the dissolution of high-fired BeO are shown in Table 7.1.[10] These data were generated by preparing a slurry of UOX (high fired BeO available from Brush Wellman, Cleveland, OH), spiking Whatman 541 filter papers with this slurry, and analyzing these filters by dissolution in 1% aqueous ABF solution. The solution volume was 5 mL, and the dissolution time was 30 min. Rotation was conducted at room temperature and at elevated temperature; the samples were not rotated or stirred. All of these experiments were conducted using standard 15 mL conical centrifuge tubes.

Table 7.1 shows that recovery from dissolution is dependent on the amount of BeO present and the temperature of dissolution. There is another study where larger BeO particles were processed using ABF solutions at 80 °C.[11] Lower recoveries were reported; however, no experimental details were given in terms of the amount of BeO to ABF solution, or whether the reaction vessel shape allowed good interaction between the solid and the liquid phases. Since a dissolution process is dependent on the particle size, the protocol needs to be optimized for the largest particle size depending on the need of the end user. In another study the dissolution method was extended to soils, sediments and fly ash.[12] These experiments showed that the dissolution rate of these materials could be increased by increasing the concentration of ABF, and increasing both temperature and time. A study on the dissolution of high-fired BeO particles about 200 μm in size was completed recently.[13] The results showed that, by using 3% ABF at 90 °C, almost 100% recovery could be achieved in a time period of 4 h when 1 mg of BeO was used in 5 mL dissolution solution and particles had good access to the liquid medium. However, it is important to remember that to reach the US Department of Energy (DOE) action level[14] only 0.2 μg of BeO needs to be solubilized in the dissolution medium. One spherical particle of BeO, which is 200 μm in diameter, has a beryllium content of 4.5 μg, which means a recovery of only 5% is needed to draw attention; however, if the intention is a true analysis, then close to 100% recovery will be needed.

7.2.2.2 Fluorescence Indicator

When the dissolution process using ABF was adopted for test development, it was clear that a fluorescent indicator that would work with this procedure must be capable of tolerating large concentrations of fluoride. The indicator, 10-hydroxybenzo[h]quinoline-7-sulfonate (10-HBQS) had previously been reported to tolerate up to 20 000 000 equivalents of fluoride; in addition, this is a water-soluble dye and chelates well with Be(II) ion[15] (Figure 7.1).

A tightly bound hydrogen bonded proton leads to weak triplet emission at 580 nm. When the proton is displaced by a metal such as beryllium, fluorescence emission is observed at 475 nm (Figure 7.2).

To keep the test process simple and to avoid titration, the dye solution was buffered with lysine monohydrochloride. This ensured that when the acidic ABF solution was mixed with the dye solution, the pH always stayed in excess of 12 – a necessary condition to bind beryllium to the 10-HBQS dye. Further, the high pH also ensured low interference from the other metals, as most of these precipitate and the binding constant of the dye to beryllium is high. Further specificity for beryllium was also achieved by adding EDTA (disodium dihydrate ethylenediamine tetraacetic acid) to the dye solution in order to bind and mask the effects from the other metals.[16,17] Iron and titanium impart a yellow color to the final solution; these tend to precipitate in a few hours and can be removed by filtration once the solutions become clear.[6,18]

Figure 7.3 shows the excitation spectrum of a measurement solution comprising one part of 1% ABF solution with different concentrations of beryllium when mixed with 19 parts of the dye solution as outlined in the standard NIOSH and ASTM International methods. The dye solution was prepared by the addition of 12.5 mL of 10.7 mM EDTA and 25 mL of 107 mM L-lysine monohydrochloride to 3 mL of 1.1 mM 10-HBQS. The pH was adjusted to 12.85 with the careful addition of 10 M NaOH, and water added to a total of 50 mL. This excitation was measured based on the emission at 475 nm and a bandwidth of ± 5 nm. The sample can be excited at any of the three peaks shown in Figure 7.2. However inexpensive instruments with excitation capabilities at 365 and 385 nm are readily available; Figure 7.3 shows the emission spectrum of the same solutions when excited by a source at 365 ± 10 nm.

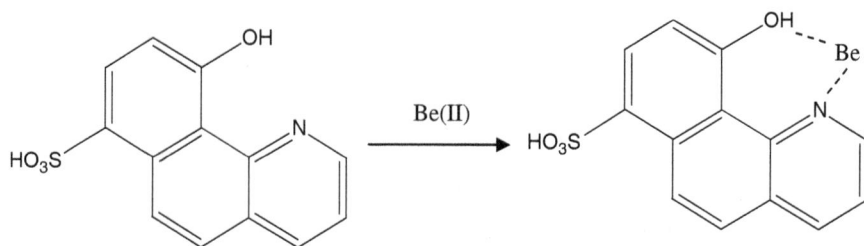

Figure 7.1 Binding of beryllium ion to 10-HBQS.

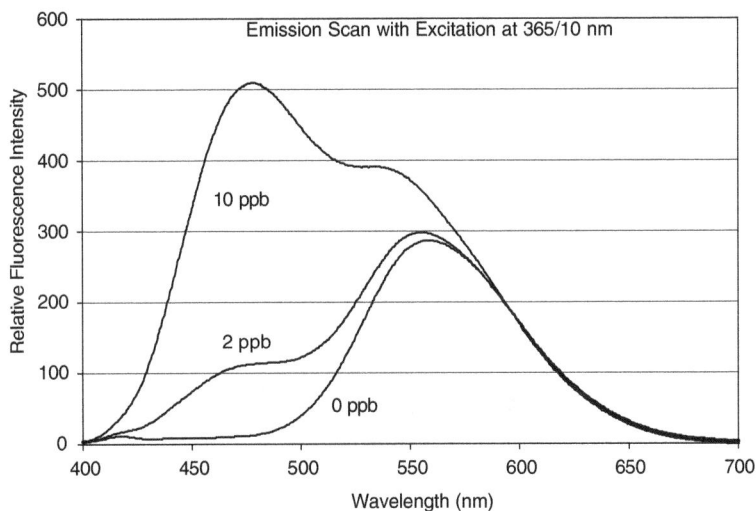

Figure 7.2 Emission spectrum of samples with various amounts of beryllium when mixed with 10-HBQS dye solution.

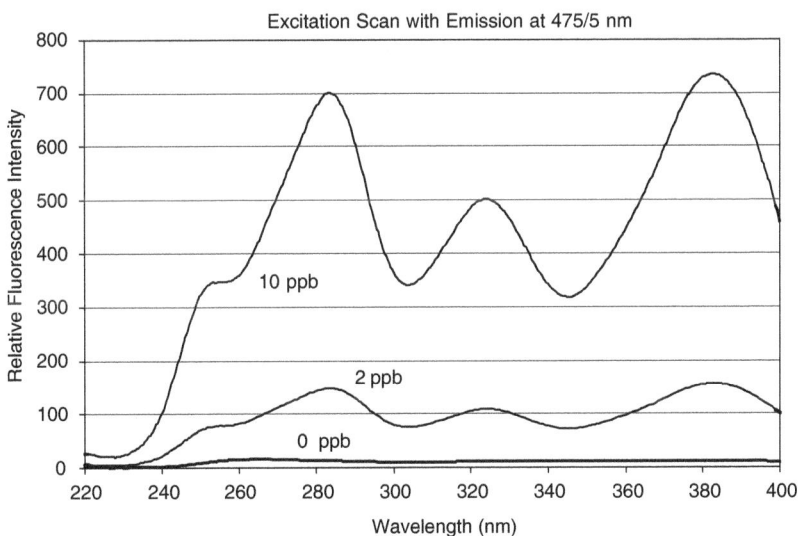

Figure 7.3 Excitation spectrum of samples with various amounts of beryllium when mixed with 10-HBQS dye solution.

In developing the standard test method, it has been demonstrated that the procedure results in good agreement when the data from different labs are compared, and, as expected, other metals do not interfere with the analysis.[6,18] To check on non-interference, some of the metals were added in concentrations in excess of 50 000 fold as compared to beryllium.

The sensitivity of the system is dependent on many factors including the ratio of the dissolution solution (with beryllium) and the dye solution. This ratio has been varied from 1 : 19 (dilution factor of 20×) to 1 : 4 (dilution factor of 5×) for 1% ABF dissolution solution. It is important to maintain pH higher than 12 to ensure good binding of the dye and beryllium.

It has been shown that beryllium in wipe or air filter samples can be quantified at 0.002 µg (2 ng), with the detection limit being even lower.[19] An amount of 0.002 µg beryllium in the wipe or air filter is equivalent to quantifying 0.08 parts per billion (ppb) of beryllium in the final measurement solution at 5× dilution.

7.2.2.3   Adaptation for Soils, Sediments, and Fly Ash

To analyze soils, sediments, and fly ash, the method described above has been slightly modified by changing the dissolution conditions.[12] The main change in dissolution condition for soils has been use of 3% ABF solution with a dissolution temperature of 90 °C and a time period of 40 h. This procedure has also been approved as an ASTM test method.[9]

Important results are shown in Figure 7.4, where they are compared with an analysis performed using ICP-MS. In these data, the Standard Reference Materials are from the Geological Survey of Japan (GSJ), the Canadian Certified Reference Materials Project (CCRMP), and the US National Institute of Standards and Technology (NIST). Samples JR-3, JA-2 and SY2 are crushed rocks, sample JB2 is volcanic ash, sample Till-1 and 2702 are soils, samples 1944 and 2710 are sediments, and sample 1633a is a fly ash.

In order to capitalize on the success of this method and the promise of being able to analyze beryllium contamination in a variety of samples, this method is being developed for high throughput analysis.

In one analytical technique, use of a flow cell has been suggested.[20] In another, tools from the biological industry are being adopted, whereby fluorescence plate readers are being used to analyze multiple samples.[21] For the latter, the analytical procedure is being integrated with an automated sample preparation method using liquid handling systems. This will allow several hundred samples to be analyzed per day and significantly reduce labor, instrumentation needs, consumables, and chemical waste.

7.3   Atomic Absorption

7.3.1   Background

Atomic absorption (AA) spectroscopy uses the absorption of light to measure the concentration of gas phase atoms. Since samples are usually liquids or solids, the analyte atoms or ions must be vaporized in a flame or graphite furnace. The atoms absorb ultraviolet or visible light, and make transitions to higher electronic energy levels. The analyte concentration is determined from the amount of absorption. Concentration measurements are usually determined

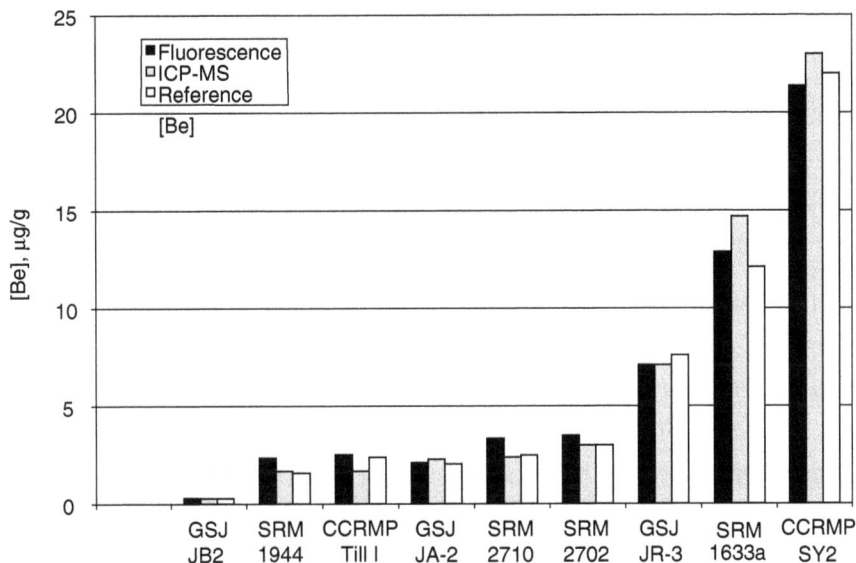

Figure 7.4 Fluorescence analysis of various standard materials and comparison with published values and ICP-MS analytical method.

from a working curve, after calibrating the instrument with standards of known concentration.

Several methods have also been reported for the atomic absorption spectrometric determination of beryllium after solid phase extraction. Okutani *et al.*[22] determined beryllium at $\mu g\,mL^{-1}$ levels in water samples by graphite furnace atomic absorption spectrometry (GFAAS) after preconcentration and separation as the beryllium acetylacetone complex on activated carbon. Szczepaniak and Szymanski[23] determined trace beryllium by GFAAS after preconcentration on silica gel with immobilized morin. Beryllium in the range $0.07-0.184\,\mu g\,mL^{-1}$ could be determined by this method. Kubova *et al.*[24] determined trace beryllium in mineral waters by GFAAS after preconcentration on a salicylate chelating resin.

7.3.2 Applications

7.3.2.1 Determination of a Trace Amount of Beryllium in Water Samples by Graphite Furnace Atomic Absorption Spectrometry after Preconcentration and Separation as a Beryllium–Acetylacetonate Complex on Activated Carbon

This is a simple preconcentration method which involves selective adsorption using activated carbon as an adsorbent and acetylacetone as a complexing agent.[25] It is used for the determination of trace amounts of beryllium by GFAAS. The beryllium–acetylacetonate complex is adsorbed onto activated carbon at pH 8–10. The activated carbon which adsorbed the beryllium–acetylacetonate complex is separated and dispersed in pure water, and the resulting suspension introduced directly into the graphite furnace atomizer. The detection limit is $0.6 \, ng \, L^{-1}$ (S/N = 3), and the relative standard deviation at $0.25 \, pg \, L^{-1}$ was 3.0–4.070 *(n = 6)*. No interference is seen from the major ions such as Na(I), K(I), Mg(II), Ca(II), Cl⁻, and SO_4^{2-} in seawater, or from other minor ions. The method was applied to the determination of nanogram per milliliter levels of beryllium in seawater and rainwater.

7.3.2.2   Solid Phase Extraction Flame Atomic Absorption Spectrometric Determination of Ultra-Trace Beryllium

This method is based on the preconcentration of Be(II) on an OH-form strong base anion exchange resin.[26] The retained beryllium is eluted with 1.5 M HCl and measured by flame atomic absorption spectrometry (AAS) at 234.9 nm. Be(II) in the range 0.05–15 µg could be determined. The relative standard deviation (RSD) for 10 replicate measurements of $20 \, ng \, mL^{-1}$ beryllium was 1.2% and the 3σ limit of detection of the method was $0.045 \, ng \, mL^{-1}$. The method was applied to the determination of beryllium in several water samples. In solutions with ionic strengths > 0.2, the adsorption of beryllium on the resin fell significantly, because passing solutions with high ionic strength causes elution of the OH⁻ from the column. Therefore, the method is less suitable for determination of beryllium in solutions with ionic strength > 0.2.

7.4   UV–Visible Spectroscopy

Ultraviolet–visible spectroscopy or ultraviolet–visible spectrophotometry (UV/VIS) involves the spectroscopy of photons in the UV–visible region. It uses light in the visible and adjacent near ultraviolet (UV) and near infrared (NIR) ranges. In this region of the electromagnetic spectrum, molecules undergo electronic transitions. This technique is complementary to fluorescence spectroscopy, in that fluorescence deals with transitions from the excited state to the ground state, while absorption measures transitions from the ground state to the excited state. This section also addresses those methods that have been developed to look at visual color changes to estimate the quantity of beryllium.

A variety of chemistries have been adopted to look at color change or optical properties in presence of beryllium. Table 7.2 summarizes a number of these methods.

None of these tests have been developed to a point where they can be adopted by regulatory agencies or approved as standardized quantitative test methods. In addition, these test procedures have not been shown to be effective for difficult to solubilze beryllium compounds such as high-fired beryllium oxide. Recently, wipes for semi-quantitative analysis have been commercialized

Table 7.2 Colorimetric methods for beryllium determination[a]

Beryllium Form/Medium Evaluated	Optical Compound	Reference
Beryllium in solution	Disodium salt of *o*-carboxy-phenylazo chromotropic acid (infra-red region)	27
Beryllium in water	Chrome Azurol S	28
Beryllium in water	Eriochrome cyanine R	29
Inorganic salts in solution	LI-complex (reaction product of 2-trichloro-methylbenz-imidazole (TCMB)pyridine)	30
Beryllium in drinking water	Aluminon	31
Beryllium determination on human skin (skin treated with H_2SO_4)	Chrome Azurol S and NH_3 buffered solution; color compared to standard	32
Beryllium determination in water and biological samples	Precipitate of beryllium ammonium phosphate with ammonium molybdate, then treated with succinyldihy-droxamic acid	33
Beryllium in coal fly ash	Chrome Azurol S in the presence of Zephiramine (ZCl)	34
Beryllium(II) determination in water and BeO in atmosphere	Anion exchange resin Amber-lite IRA-400 and Chrome Azurol S in aqueous solution	35
Beryllium determination in water and waste water	Chromazurine S and hexa-decylpyridine chloride (pH 5–5.3)	36
Absorption spectra of complexes with alkali and alkaline earth metal ions, including beryllium	4-(2,4-Dinitrophenylazo)-phe-nol(I), (II), and (III)	37
Beryllium determination in water	Ion exchange colorimetry with Eriochrome R	38
Beryllium determination in ore samples	Beryllon(III)	39
Beryllium determination in water	8-Hydroxynaphthalene-3,6-disulfonic acid (1-azo-1)-2,4-resorcinol	40
Analysis of various metals in solution	Isticin-9-imine, alizarin-9-imine, 3-sulfoalizarin-9-imine	41
Spectrophotometric determination of fluoride	Beryllium–carboxylate dye complex	42
Beryllium determination in air	Chrome Azurol S	43
Be(II) determination in solution	$K_3Cr(CN)_6$	44
Beryllium surface spot test	Chrome Azurol S	45
Beryllium chloride in solution	Ammonium aurintricarboxy-late (aluminon)	46
Beryllium determination in water	Sample extracted three times with CCl_4 (pH adjusted to 5–9), $Na_2H_2[(O_2CCH_2)2CH_2]_2(I)$ and Ac_2CH_2 and beryllon added	47
Beryllium determination in waste waters	Beryllon(II)	48

Table 7.2   (*Continued*)

Beryllium Form/Medium Evaluated	Optical Compound	Reference
Beryllium determination in bronze sample	Eriochrome cyanine	49
Beryllium determination in coal ash	Beryllon(II)	50
Beryllium determination in solution	Chrome Azurol S	51
Beryllium determination in solution	Beryllon(II), thoron, and arsenazo(I) compared	52
Beryllium determination in bronze sample	Chrome Azurol S	53
Beryllium on surfaces and air	Chrome Azurol S	54
Beryllium on surfaces	Chrome Azurol S	55
Beryllium in water, beryl ore, and beryllium alloy	Anthralin at pH 11.3	56

[a]This table has largely been adopted from Taylor and Sauer[54] and updated.

using Chrome Azurol S which changes color based on the beryllium content.[56] These wipes have been shown to change color at 0.2 μg of beryllium in order to meet the US DOE action limit.

7.5   Electrochemistry

There are a few electrochemical methods available in the literature for the determination of beryllium, but these are not widely used. There are only limited reports in the literature on Be(II) selective electrodes to directly determine the beryllium concentration.

7.5.1   Adsorptive Stripping Voltammetric Measurements of Trace Beryllium at the Mercury Film Electrode

This method involves adsorptive stripping voltammetric measurement of trace beryllium using preconcentration by adsorption of a beryllium–arsenazo-I complex at a preplated mercury-coated carbon fiber electrode.[57] It is a sensitive cathodic stripping protocol for detecting trace beryllium based on the adsorptive accumulation of the Be–arsenazo-I complex at a mercury film electrode. Optimal conditions were found to be a 0.05 M ammonium buffer (pH 9.7) containing 5 μM arsenazo-I, an accumulation potential of 0.0V (*vs.* Ag/AgCl) and a square-wave voltammetric scan. A linear response is observed over the 10–60 μg L^{-1} concentration range (90 s accumulation), along with a detection limit of 0.25 μg L^{-1} beryllium with a 10 min accumulation. The same mercury film can be used for a prolonged operation with proper electrochemical cleaning. High stability is seen from the reproducible response of a 100 μg L^{-1} beryllium solution over 2.5 h operation. Examples of data using different stripping modes are shown in Figure 7.5. The new sensor shows

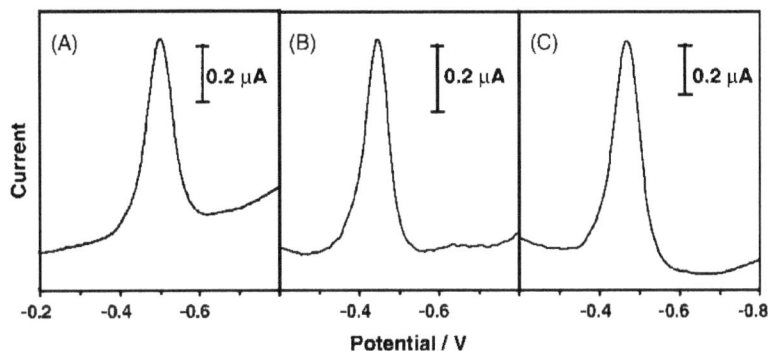

Figure 7.5 Comparison of different stripping modes for a mercury-coated carbon-fiber electrode; supporting electrolyte, 0.05M ammonium buffer (pH 9.7) containing 100 _gl–1 beryllium and 5 _M arsenazo-I: (A) Linear scan voltammetry; (B) differential pulse voltammetry; (C) square-wave voltammetry. From ref. 57.

promise for on-site environmental and industrial monitoring of beryllium. This procedure obviates the need for the large mercury-drop electrode and related mercury disposal issues. The same preplated mercury film can be employed for multiple measurements of beryllium.

7.5.2 Beryllium-Selective Membrane Electrode Based on Benzo-9-crown-3

A polyvinylchloride (PVC) membrane electrode for Be(II) ions based on benzo-9-crown-3 as membrane carrier was prepared and tested.[58] The sensor exhibited a Nernstian response for Be(II) ions over a wide concentration range $(4.0 \times 10^{-3}$–2.5×10^{-6} M) with a limit of detection of 1.0×10^{-6} M $(9.0 \times 10^{-3}$ parts per million). The sensor has a response time of 50 s and can be used for several months without any divergence in potential. The electrode displayed good selectivity for Be(II) over a wide variety of other cations including alkali, alkaline earth, transition, and heavy metal ions, and could be used in a pH range of 2.0–6.0. It was successfully applied to the determination of beryllium in a mineral sample.

Benzo-9-crown-3 (B9C3) was found to be an excellent neutral carrier in construction of a Be(II) PVC membrane electrode. The small size of the B9C3 cavity increased both the stability and the selectivity of its beryllium complex over those of other metal ions. In addition, the existence of a benzo ring on the crown's ring results in its diminished solubility in aqueous solutions.

The Be(II) ion-selective electrode was found to work well under the laboratory conditions. It was successfully applied to the determination of beryllium in mineral samples such as beryl. The beryllium content in a prepared solution was determined using the proposed membrane sensor and atomic absorption spectrometry. The results obtained by the ion selective electrode are in satisfactory agreement with those obtained by AA.

7.5.3 New Diamino Compound as Neutral Ionophore for Highly Selective and Sensitive PVC Membrane Electrode for Be(II) Ion

This method uses diamine 2,6-bis[2-(*o*-aminophenoxy)methyl]-4-bromo-1-methoxybenzene compound as a beryllium ion carrier in a PVC membrane electrode for potentiometric determination of Be(II). [59] The electrode exhibited a Nernstian response to Be(II) ion over a wide concentration range from 3.0×10^{-6} to 7.0×10^{-2} M, and a detection limit of 2.0×10^{-6} M. It had an appropriate response time and suitable reproducibility, and could be used for a period of at least one month without degradation of performance. The potentiometric response is independent of the pH of the test solution in the range of 4.0–7.0. The sensor revealed good selectivity toward Be(II) ion with respect to many alkali, alkaline earth, transition, and heavy metal ions. It was applied successfully to the determination of beryllium in tap water samples and also samples containing interfering ions.

7.5.4 Beryllium-Selective Membrane Sensor Based on 3,4-Di[2-(2-Tetrahydro-2H-Pyranoxy)]Ethoxy Styrene–Styrene Copolymer

3,4-Di[2-(2-tetrahydro-2H-pyranoxy)]ethoxy styrene–styrene copolymer was used to prepare a beryllium-selective PVC-based membrane electrode. [60] The resulting sensor works well over a wide concentration range (1.0×10^{-6} to 1.0×10^{-3} M) with a Nernstian slope of 29 mV per decade of Be(II) activity over a pH range 4.0–8.0. The detection limit of the electrode is 8.0×10^{-7} M (7.6 ng mL^{-1}). The electrode showed excellent selectivity toward Be(II) ion with regard to alkali, alkaline earth, transition, and heavy metal ions. It was successfully applied to the determination of beryllium in a mineral sample.

The Be(II) ion-selective electrode worked well under laboratory conditions. It was applied to the direct measurement of Be(II) in beryl samples after suitable treatment of the mineral. The results obtained by the ion-selective electrode are in satisfactory agreement with those obtained by AA.

7.5.5 New Diamino Compound as Neutral Ionophore for Highly Selective and Sensitive PVC Membrane Electrode for Be(II) Ion

In this work, five macrocyclic diamides were investigated to characterize their ability as beryllium ion carriers in potentiometric PVC membrane electrodes. [61] The electrodes based on 1,15-diaza-3,4;12,13-dibenzo-5,8,11-trioxabicyclo[13,2,2] heptadecane-2,14-dione exhibited a Nernstian response for Be(II) ion over wide concentration ranges [from 3.0×10^{-6} to 3.0×10^{-2} M for polymeric membrane electrode (PME), and from 5.0×10^{-7} to 2.0×10^{-2} M for coated glassy carbon electrode (CGCE), and very low detection limits (2.0×10^{-6} M for PME and 4.0×10^{-7} M for CGCE]. The electrodes had low resistances, fast responses, satisfactory reproducibilities and, good selectivities for Be(II). The potentio-metric response of the electrodes is independent of the pH of the test solution in the pH range 4.0–7.5. The sensors were used to determine beryllium ion in water samples.

7.6 Other Methods

7.6.1 Utilization of Solid Phase Spectrophotometry for Determination of Trace Amounts of Beryllium in Natural Water

Solid-phase spectrophotometry (SPS) is a technique based on the pre-concentration of the substance in question on a solid using complexing or other reagents, followed by measurement of the absorbance of the species in the solid phase.[62] This provides SPS with an increase in selectivity and sensitivity over other methods. Detection limits as low as $0.09 \, \mu g \, L^{-1}$ have been reported. EDTA is used as the complexing agent in solution for the spectrophotometric method of beryllium determination.

In this work, a dextrane-type exchanger (*i.e.* a mixed ligand complex) is used as the basis of a method to determine beryllium with aurintricarboxylic acid (ATCA) as complexing agent. Beryllium reacts with ATCA in the presence of EDTA to give a complex with a high molar absorptivity ($1.50 \times 10^7 \, L \, mol^{-1}$ cm^{-1}), which is fixed on a dextran-type anion-exchange gel (Sephadex DEAE A-25). The absorbance of the gel, at 575 and 750 nm, packed in a 1.0 mm cell, is measured directly. Calibration is linear over the range $0.03–1.0 \, \mu g \, L^{-1}$ with RSD of $<2.4\%$ ($n = 8.0$). The detection and quantification limits of the 500 mL sample method are 6.0 and $20 \, ng \, L^{-1}$, respectively, using a 90 mg exchanger. For 1000 mL samples, the detection and quantification limits are 5.0 and $17 \, ng \, L^{-1}$, respectively, using a 45 mg exchanger. Increasing the sample volume can enhance the sensitivity. The methods were applied to the determination of beryllium in tap, mineral and well water using the standard addition technique with recoveries of close to 100% at concentration levels of 0.032, 0.221 and $10.92 \, g \, L^{-1}$, respectively.

7.6.2 Selective Determination of Beryllium(II) Ion at Picomole per Decimeter Cubed Levels by Kinetic Differentiation Mode Reversed-Phase High-Performance Liquid Chromatography with Fluorometric Detection Using 2-(2′-Hydroxyphenyl)-10-hydroxybenzo[h]quinoline as Precolumn Chelating Reagent

This method for the determination of the Be(II) ion uses reversed-phase high-performance liquid chromatography (HPLC) with fluorometric detection using 2-(2′-hydroxyphenyl)-10-hydroxybenzo[h]quinoline (HPHBQ) as a precolumn (off-line) chelating reagent.[63] The reagent HPHBQ forms the kinetically inert

Be chelate compatible with high fluorescence yield, which is appropriate to the HPLC-fluorometric detection system. The Be-HPHBQ chelate is efficiently separated on a LiChrospher 100 RP-18(e) column with a methanol (58.3 wt%)-water eluent containing $20\,mmol\,kg^{-1}$ of tartaric acid and is fluorometrically detected at 520 nm with the excitation at 420 nm. Under the conditions used, the concentration range of $20–8000\,pmol\,dm^{-3}$ of Be(II) ion can be determined without interferences from $10\,mmol\,dm^{-3}$ each of common metal ions, typically Al(III), Cu(II), Fe(III), and Zn(II), and still more coexistence of Ca(II) and Mg(II) ions at $0.50\,mmol\,dm^{-3}$ and $5.0\,mmol\,dm^{-3}$, respectively, is tolerated. The detection limit is $4.3\,pmol\,dm^{-3}$ $(39\,fg\,cm^{-3})$.

References

1. H. A. Laitinen and P. Kivalo, *Anal. Chem.*, 1952, **24**, 1467–1471.
2. F. Capitan, E. Manzano, A. Navalon, J. L. Vilchez and L. F. Capitan-Vallvey, *Analyst*, 1989, **8**, 969–973.
3. L. A. Saari and W. R. Seitz, *Analyst*, 1984, **109**, 655–657.
4. B. K. Pal and K. Baksi, *Microchim. Acta*, 1992, **108**, 275–283.
5. K. Morisige, *Anal. Chim. Acta*, 1974, **73**, 245–254.
6. E. M. Minogue, D. S. Ehler, A. K. Burrell, T. M. McCleskey and T. P. Taylor, *J. ASTM Int.*, 2005, **2**, [JAI13168].
7. ASTM D7202, *Standard Test Method for the Determination of Beryllium in the Workplace using Field-based Extraction and Fluorescence Detection*, ASTM International, West Conshohocken, PA, 2006.
8. 'Method 7704. Beryllium in air by field portable fluorometry' and 'Method 9110. Beryllium in surface wipes by field portable fluorometry', in *NIOSH Manual of Analytical Methods*, ed. P. C. Schlecht and P. F. O'Connor, National Institute for Occupational Safety and Health, Cincinnati, OH, 4th edn, 1994–2006, www.cdc.gov/niosh/nmam/, accessed 12 February 2009.
9. ASTM D7458, *Standard Test Method for Determination of Beryllium in Soil, Rock, Sediment, and Fly Ash Using Ammonium Bifluoride Extraction and Fluorescence Detection*. ASTM International, West Conshohocken, PA, USA, 2008.
10. T. M. McCleskey, A. Agrawal and K. Ashley, Backup data – method nos. 7704 and 9110, Beryllium, Issue 1, NIOSH Docket Office, Cincinnati, OH, 2007.
11. A. B. Stefaniak, G. C. Turk, R. M. Dickerson and M. D. Hoover, *Anal. Bioanal. Chem.*, 2008, **391**, 2071–2077.
12. A. Agrawal, J. Cronin, A. Agrawal, J. Tonazzi, K. Ashley, M. Brisson, B. Duran, G. Whitney, A. Burrell, T. McCleskey, J. Robbins and K. White, *Env. Sci. Technol.*, 2008, **42**, 2066–2071.
13. M. Goldcamp and K. Ashley, private communication, to be published in 2009, NIOSH, Cincinnati, OH, USA.

14. US Code of Federal Regulations, 10 CFR Part 850, *Fed. Regist.*, 1999, 64(8th December), 68854–68914.
15. H. Matsumiya, H. Hoshino and T. Yotsuyanagi, *Analyst*, 2001, **126**, 2082–2086.
16. NIST Standard Reference Database 46, NIST Critically Selected Stability Constants of Metal Complexes: Version 8.0, US National Institute of Standards and Technology, Gaithersburg, MD, www.nist.gov/srd/nist46.htm, accessed 12 February 2009.
17. D. B. Do Nascimento and G. Schwedt, *Anal. Chim. Acta*, 1993, **283**, 909–915.
18. A. Agrawal, J. Cronin, J. Tonazzi, T. M. McCleskey, D. S. Ehler, E. M. Minogue, G. Whitney, C. Brink, A. K. Burrell, B. Warner, M. J. Goldcamp, P. C. Schlect, P. Sonthalia and K. Ashley, *J. Environ. Monit.*, 2006, **8**, 619–624.
19. K. Ashley, A. Agrawal, J. Cronin, J. Tonazzi, T. M. McCleskey, A. K. Burrell and D. S. Ehler, *Anal. Chim. Acta*, 2007, **584**, 281–286.
20. W. A. Spencer and A. A. Ekechukwu, Specification for Automation of ASTM D7202-05 for Analysis of Air Filters and Surface Wipes for Beryllium in the Workplace, WSRC-STI-2006-00135, 30 September 2006.
21. A. Agrawal and Berylliant Inc., *Automation of Fluorescence Method for Beryllium Analysis*, presented at Beryllium Health and Safety Committee, semiannual meeting, Aberdeen, MD, April 2008.
22. T. Okutani, Y. Tsuruta and A. Sakuragawa, *Anal. Chem*, 1993, **65**, 1273.
23. W. Szczepaniak and A. Szymanski, *Chem. Anal.*, 1996, **41**, 193.
24. J. Kubova, V. Nevoral and V. Stresko, *Fresenius J. Anal. Chem.*, 1994, **348**, 287.
25. T. Okutani, Y. Tsuruta and A. Sakuragawa, *Anal. Chem.*, 1993, **65**, 1273–1276.
26. A. Afkhami, T. Madrakian, A. A. Assl and A. A. Sehhat, *Anal. Chim. Acta*, 2001, **437**, 17–22.
27. A. K. Majumdar and C. P. Savariar, *Fresenius Z. Anal. Chem.*, 1960, **176**, 170–174.
28. H. Dong, M. Jang, G. Zhao and M. Wang, *Anal. Sci.*, 1991, **7**, 69–72.
29. N. Ershova and V. Ivaonx, *Fresenius Z. Anal. Chem.*, 2001, **371**, 556–558.
30. A. Kielczewska and L. Konopski, *Pol. Organika*, 1996, **1995**, 33.
31. D.A. Lytle, G.K. George and J.U. Doerger, in *Proceedings of the Water Quality Technology Conference*, American Water Works Association, Denver, CO, 1992, Part I, p. 683.
32. N.I. Egorova, D.B. Slobodin, T.K. Strygina and T.N. Pavlenko, Russian Patent AN 250276, 1992.
33. L. Cherian and V. K. Gupta, *Asian Environ.*, 1990, **12**, 27.
34. H. Nishida and H. Bunseki, *Kagaku*, 1990, **39**, 805.
35. A. Chiba and T. Ogawa, *Dep. Mater. Chem.*, 1988, **56**, 627.
36. X. Qui, J. Chen and Y. Zhy, *Huanjing Kexue*, 1988, **9**, 55.

37. K. Nakashima, S. Nakatsuji, S. Akiyama, T. Kaneda and S. Misumi, *Chem. Pharm. Bull.*, 1986, **34**, 168.
38. R. Yi and R. Wang, *Fenxi Huaxue*, 1985, **13**, 130.
39. A. Zhao and H. He, *Fen Hsi Hua Hsueh*, 1981, **9**, 246.
40. T. Chen, J. Gao and L. Kong, *Fen Hsi Hua Hsueh*, 1981, **9**, 56.
41. M. Blanco and J. Barbosa, *Talanta*, 1980, **27**, 371.
42. J. Ramasamy and J. L. Lambert, *Anal. Chem.*, 1979, **51**, 2044.
43. H. R. Mulwani and R. M. Sathe, *Analyst*, 1977, **102**, 137.
44. W. U. Malik and K. D. Sharma, *Ind. J. Chem.*, 1975, **13**, 1232.
45. E. Mordberg and E. M. Fil'kova, *Gig. Sanit.*, 1974, **6**, 71.
46. B. K. Avinashi and S. K. Banerji, *Ind. J. Chem.*, 1972, **10**, 213.
47. T. G. Kornienko and A. I. Samchuk, *Ukr. Khim. Zh.*, 1972, **38**, 917.
48. O. V. Yanter and E. A. Orlova, *Ref. Zh. Khim.*, 1970, **2** (Abstract 18G169), 89.
49. K. Kasiura, *Chem. Anal.*, 1971, **16**, 407.
50. R. F. Abernathy and E. A. Hattman, *US Govt. Res. Dev. Rep.*, 1970, **71**, 60.
51. L. Sommer and V. Kubáò, *Anal. Chim. Acta*, 1969, **44**, 333.
52. D. I. Eristavi, V. D. Eristavi and S. h. A. Kekeliya, *Tr. Gruz. Politekh, Inst.*, 1968, **5**, 56.
53. E. Cioaca, *Cercet. Met.*, 1967, **9**, 693.
54. T. Taylor and N. Sauer, *J. Hazard. Mater.*, 2002, **93**, 271–283.
55. T. M. Tekleab, G. M. Mihaylov and K. S. Kirollos, *J. Environ. Monit.*, 2006, **8**, 625–629.
56. A. Beiraghi and S. Babaee, *J. Iran. Chem. Soc.*, 2007, **4**, 459–466.
57. J. Wang, S. Thongngamdee and D. Lu, *Anal. Chim. Acta*, 2006, **564**, 248–252.
58. M. R. Ganjali, A. Moghimi and M. Shamsipur, *Anal. Chem.*, 1998, **70**, 5259–5263.
59. A. Soleymanpour, N. A. Rada and K. Niknamb, *Sens. Actuators B*, 2006, **114**, 740–746.
60. M. Shamsipur, M. R. Ganjali, A. Rouhollahi and A. Moghimid, *Anal. Chim. Acta*, 2001, **434**, 23–27.
61. M. Shamsipur, A. Soleymanpour, M. Akhond and H. Sharghi, *Electroanalysis*, 2004, **16**, 282–288.
62. A. S. Amin, *Anal. Chim. Acta*, 2001, **437**, 265–272.
63. H. Matsumiya and H. Hoshino, *Anal. Chem.*, 2003, **75**, 413–419.

CHAPTER 8

Data Use, Quality, Reporting, and Communication

NANCY E. GRAMS[a] AND CHARLES B. DAVIS[b]

[a] Advanced Earth Technologies, 40w840 Bowes Bend Drive, Elgin, IL 60124, USA; [b] EnviroStat, 3468 Misty Court, Las Vegas, NV 89120, USA

Abstract

After a general discussion of laboratories, testing data, and reports, this chapter focuses on three topics: "limit" values encountered with laboratory data; Data Quality Objectives (DQOs) and Measurement Quality Objectives (MQOs) for selection of laboratories, methods, and data requirements; and the benefits and issues associated with using uncensored data.

Data censoring and the various limit values in use are traced to statistically defined concepts of critical, detectable, and quantifiable levels set down by Lloyd Currie in 1968. These are discussed in detail using ICP-AES beryllium measurements to illustrate their conceptual and technical subtleties. The current U.S. EPA Method Detection Limit (MDL) receives particular attention due to its pre-eminent position in U.S. regulatory affairs and the legal challenges it currently faces. Recent technical advances, notably those due to ASTM, are also discussed.

DQOs and MQOs involve deliberate quantitative planning of an investigation. These are discussed in terms of the statistical properties required of laboratory measurements, again using simple conceptual examples, for decisions based on individual measurements. DQOs and MQOs for decision-making using entire datasets are also discussed, including whether to use censored or uncensored data and issues involving the actual data distributions that may be encountered.

Beryllium: Environmental Analysis and Monitoring
Edited by Michael J. Brisson and Amy A. Ekechukwu
© Royal Society of Chemistry 2009
Published by the Royal Society of Chemistry, www.rsc.org

Finally, technical aspects of the use of uncensored data is discussed, as are several non-technical issues related to public relations, risk communication and accreditation requirements.

8.1   Introduction and Overview

Beryllium measurements are used to make decisions in a variety of settings, often related to the protection of the health of workers, other potentially affected persons, or the environment, and most often driven by government regulation or industry standards. Analytical results are generally provided to the user in an analytical report. The data user must interpret the report (perhaps making a few inquiries of the laboratory about data qualifications or case narratives), integrate testing data with field and sampling information, and then synthesize this input, making decisions about the impact of beryllium on human health and/or the environment. As previous chapters have described, arriving at the point where beryllium monitoring information can be applied to a decision requires sequential efforts by several disciplines. Laboratories become the focal point as the final step in the process, with the laboratory's report being the primary communication to the data user.

Much information is filtered and condensed by the laboratory into that report. From "raw" sample management and instrument data – including calibration data, sample preparation information, results of quality control tests, and actual sample data – a simplified product is produced. Laboratories are also responsible for incorporating and utilizing project-specific information provided by project planners such as project-specific data quality objectives, non-standard quality control tests and criteria, as well as chain-of-custody and field information. Additionally, laboratories are responsible to accrediting agencies for maintaining established standards of quality and defensibility through the consistent application of written quality assurance procedures.

The laboratory's report is a succinct work product summarizing this expansive, detailed, technical, and integrated effort. This chapter highlights items of particular interest in the "information-distillation" process, reviewing vital knowledge for interpretation of reported results and the technical shorthand included within. Ultimately, our purpose is to encourage a more informed understanding of various concepts encountered in analytical reports, assisting data users in initiating better communications with their laboratories as early as the project planning stage. Doing so will enhance the suitability and usability of beryllium data for the purposes and needs of individual projects.

8.1.1   Laboratory Reports

The International Organization for Standardization (ISO) has developed a minimum standard for the content of an analytical report in ISO Standard 17025.[1] (The term "testing" in the title of this standard is used quite generally for the process of obtaining laboratory or field measurements.) This standard is widely adopted by accrediting agencies and laboratories around the world. This minimum standard includes various administrative and documentation requirements that provide for traceability and legal defensibility, and also includes minimum standard "meta-data" requirements, such as the information accompanying each result (*e.g.* units of measure, test method used, identification of the sample tested, *etc.*)

Laboratories frequently provide additional information beyond the minimum as well as work products beyond paper (or electronic image) test reports. Providing a foundational understanding of this body of available work products is beyond the scope of this chapter, but the data user should understand that "packages" of detailed laboratory data appropriate for independent, third-party validation of the laboratory's testing, data reduction and reporting process can generally be provided, if specified in advance. Similarly, extended quality control (QC) reports that go beyond sample test results to include QC test results for method blanks, matrix spikes, duplicates, and second-source standard spikes (known as laboratory control samples – LCS), obtained in conjunction with the reported samples, can be provided on request.

Laboratories are also routinely asked to provide electronic data deliverables (EDDs), either instead of a paper or image report, or in addition to it. These EDDs may contain the same information provided on standardized reports of testing or may vary significantly. The most frequently requested EDDs are spreadsheet-based files in which the data can be manipulated by the user or which can easily be transferred into databases. Other information – certifications, performance testing data, quality program metrics and manuals, methods, standard operating procedures for non-method processes, quality control criteria, quality control charts, and detection limit study information, to name just a few – are routinely used and available in laboratories, as required and evaluated by accreditation bodies, and may be valuable to sophisticated users of analytical services in more fully understanding the qualities of reported data.

8.1.2 "Reporting Limits" and "Detection Limits"

Measurement systems typically used for beryllium analysis produce a response for every sample. This is in contrast to analytical systems such as gas chromatography/mass spectrometry (GC/MS) for organic constituents, where qualitative identification criteria and/or data smoothing routines censor the signal prior to assessment of the analytical response, such that a numerical result is not obtained unless the identification criteria are met. This is important to understand because, where the instrument signal is uncensored and each analytical result produces a response, it is quite possible to obtain negative beryllium values. In fact, when zero beryllium is introduced into a beryllium testing system, it would be expected that, in the absence of *bias*, the process would produce approximately 50% negative and 50% positive results.

Table 8.1   Currie's Concepts.

	Currie's Term	*Alias*	*Concept*
L_C	Decision Limit	Critical Level	*Signal* above which one declares the analyte detected
L_D	*a priori* Detection Limit	Detectable Level	*Concentration* above which one will reliably detect the analyte, using L_C for making detection decisions
L_Q	Determination Limit	Quantifiable Level	*Concentration* above which the analyte will be measured with acceptable precision

(Bias is the difference between the average measurement value and the true concentration.) An unsophisticated user might interpret a low but positive value as significant and "real", even when it is in a range of measured results so low as to be not individually reliable and of no practical significance; the system itself does not have the sensitivity to distinguish a measurement of a very small amount of beryllium from a measurement of a sample containing no beryllium. Laboratories must have a mechanism for communicating this limitation and the minimum levels at which acceptable reliability is reached, as well as (for many uses, users, and accreditation requirements) a mechanism for censoring the unreliable data. From these needs has come the use of "limit" values in analytical reports.

A huge variety of acronyms appears in test reports to represent reporting and censoring limits, including LD, LOD, DL, IDL, LQ, LOQ, LC, QL, PQL, EQL, ML, MDL, IDE, and RL. The variety in definition of these terms is multiplied by the different procedures and criteria used for their calculation, resulting in serious confusion for the data user. However, underlying these many terms and their varied implementations in laboratories is a reasonably universal conceptual framework that is useful as a basis for understanding, and as a starting point for deciphering a particular limit used in a particular report. These concepts, presented in Table 8.1, were extracted in 1968 by Lloyd Currie[2] from common practice at the time.

Most limits in use today are, at least superficially, related to Currie's statistically based theory and the three concepts developed within the theory. Currie's Concepts are discussed in some detail in Section 8.2.1; there are, however, subtleties involved that are routinely misunderstood and misapplied. Moreover, converting from theory to application presents significant technical challenges and practical compromises with statistical rigor. Data users need to understand Currie's Concepts in order to relate terms used in any laboratory back to this common basis. The user can then properly evaluate reported results in relationship to the applied limit values so that the qualities of Be measurement data above and below a reported limit may be understood and these data may be used appropriately.

8.1.3 Uses of Beryllium Data

The Data Reporting Task Force (DRTF) commissioned by the Beryllium Health and Safety Committee (BHSC) in 2006 brainstormed ways in which beryllium data are used in US Department of Energy investigations. This group characterised data uses in several ways, including:

- Sample media – air, surface, or bulk
- Purpose – exposure, control, facility assessment, or housekeeping assessment, or release of items to the public
- Degree of urgency – from routine to event response.

In particular, this group identified a most important dichotomy between decisions based on *single* beryllium test results and decisions based on *many* beryllium test results. For example, determining whether or not an entire facility satisfies some regulatory criterion of cleanliness with respect to surface concentrations of removable beryllium requires many wipe samples, comparing the entire dataset's analytical results with a regulatory criterion or standard, and typically using some type of statistical analysis. Alternatively, when assessing an individual's exposure to beryllium in the air, a single value (obtained through analysis of a filter from the individual's personal air monitor) is compared with the maximum allowable (or perhaps a lower "sentinel" value) under a specific regulation or standard. In either situation, the decision is likely to have important consequences, and an incorrect decision (based on faulty monitoring results or even on inescapable sampling or analytical variability) should be avoided to the extent possible. Accordingly, one would like to control the *rate* or probability of using an erroneous test result as the basis for the determination (falsely high or falsely low result relative to the unknown real value).

It is important to observe that the roles of Currie's Concepts (see the rightmost column of Table 8.1) are defined in terms of evaluations of individual measurements, not of entire datasets. Laboratory use of derived limit values is likewise aligned with this single-value-at-a-time use. Currie's Concepts, derived specific laboratory reporting limits, and data censoring were not invented with statistical evaluations and decisions based on entire datasets in mind. Therefore, the use of laboratory limits and the laboratory practice of data censoring should be re-evaluated (if not discarded altogether) for these decisions about the uses of datasets. However, laboratories must conform to accreditation and other regulatory requirements in providing data, qualifying and censoring data in accord with specific accreditation requirements, and/or the laboratory's written systems and programs which have been accredited. The accreditation programs of the American Industrial Hygiene Association (AIHA), for example, includes such requirements.[3]

The situation is further complicated when particular results serve multiple users, uses, and decisions. For example, breathing space measurements may be used one at a time for evaluating individuals' exposures during single work shifts, but may then later be assembled into a larger dataset used for the purpose of determining whether the facility's contamination control systems are functioning as desired. In such cases, a standard report with low data values censored may be most appropriate for the individuals' exposure assessments, while an EDD containing uncensored data that can be manipulated by work area or time period would better accommodate the multiple data (statistical) decisions. Where uncensored data are provided by laboratories to outside users, laboratories generally continue to include reporting limit meta-data so as to completely and properly represent their testing, and to address legal and accreditation requirements. Laboratories may also include disclaimers, data qualifiers, and other language as regards validity of data below their standard reporting limits.

8.2 "Detection Limits" and Related Concepts

To begin our discussion of the reporting and data censoring limit concepts, we first observe that Currie did not define Reporting Limit (RL). Unlike other limits, which are generally related to Currie's Concepts, there is no consistent concept or definition associated with RL directly related to data variability or uncertainty. Rather, a RL is a censoring point established by contract or agreement between a laboratory and the data user; in the absence of an agreement, a laboratory will use its customary RL, as established by its written quality system. Section 8.2.4 elaborates on RLs once a foundation in "detection limits" and related concepts has been established.

8.2.1 Currie's Detection and Quantitation Concepts

This section uses a single beryllium dataset to illustrate Currie's Concepts. The discussion is intentionally simplistic and conceptual. A number of technical issues that arise in attempting to implement those concepts in a satisfactory and sound manner are also identified. It is important to recognize that Currie's Concepts are rarely used as originally intended, even in the applications to radiological analyses, which were his focus at the time.

Figure 8.1 displays the example dataset. There are 100 independent samples made of GhostWipes™ spiked with amounts from 0 to 0.3 µg of beryllium (in the form of a standard solution of beryllium metal dissolved in nitric acid), provided to a single laboratory single-blind (as performance evaluation or PE samples), and tested using inductively coupled plasma atomic emission spectroscopy (ICP-AES). The horizontal axis shows the as-made concentrations (C); the vertical axis shows the measured concentrations.

Figure 8.1 also presents the fitted (weighted-least-squares) response line and the fitted standard deviation (both functions of C) in addition to the data. There is a statistically significant increase in standard deviation (StDev) as C increases. This relationship is not peculiar to this dataset, or to beryllium; rather it is to be expected with much analytical data. Ignoring the non-constant

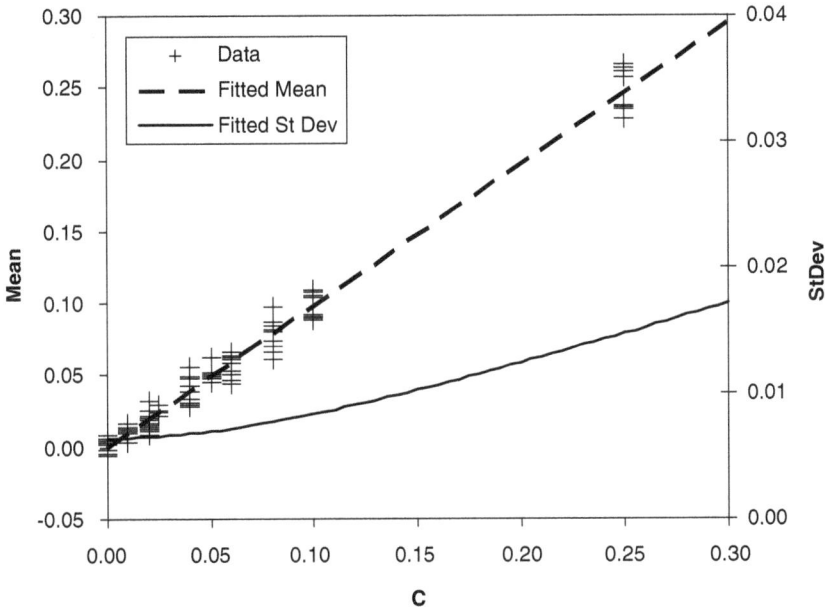

Figure 8.1 PE study data.

relationship between variability and actual concentration when developing limit values, or using limit values that assume a constant standard deviation, may be a serious error, depending on the magnitude of the changing relationship. On the other hand, taking account of it rigorously is a significant undertaking. There are a number of ways to account for this relationship, all of which add complexity to the study design and calculations. The added costs for multi-concentration studies as well as the technical complexity, along with the pre-eminent position that the Method Detection Limit (MDL), as defined by the US Environmental Protection Agency (EPA) MDL has acquired (see Section 8.2.3), have nearly universally (within the United States) resulted in laboratories implementing procedures that assume that the standard deviation is constant and employ single concentration study designs. It must be noted, however, that that those added costs are relatively much smaller with beryllium analyses using ICP-AES or inductively coupled plasma mass spectrometry (ICP-MS) than with organics analyses using GC/MS. Laboratories and users seeking more rigorous approaches to determination of limit values for beryllium should not be dissuaded by these barriers. Comprehensive statistical procedures and even software to automate the calculations have been developed (see Section 8.2.3 for more information).

With the multi-concentration design in our example, Figure 8.1 graphically depicts the significantly non-constant standard deviation, which has been accounted for using the Rocke and Lorenzato[4] model (aka the general analytical model, or the hybrid model in ASTM D6091-03[5]). In Figure 8.1, data and

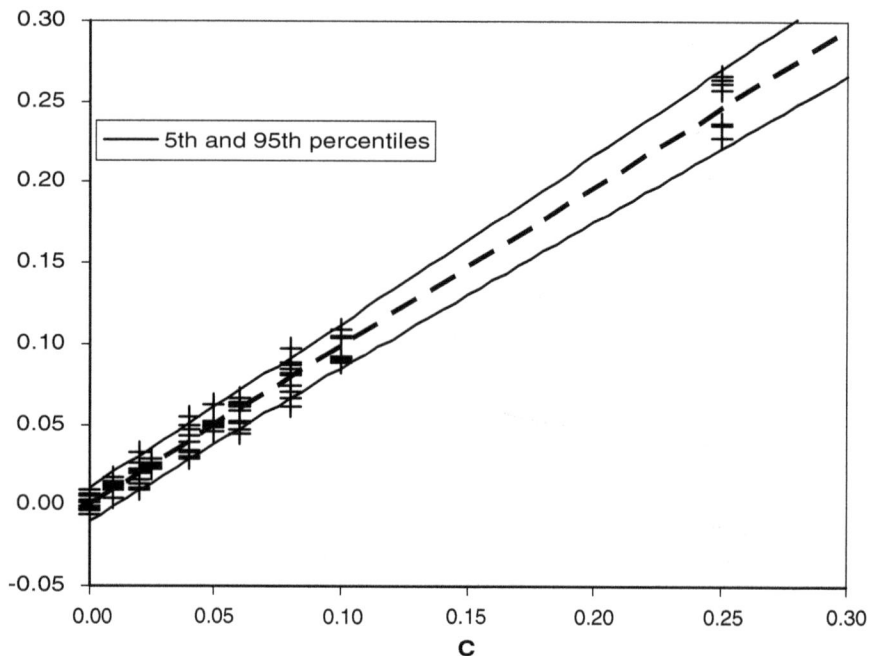

Figure 8.2 PE study data.

mean response function use the left vertical scale, while the modeled standard deviation function uses the right scale. Note in particular that, for C=0, the mean response is approximately zero, but the standard deviation is clearly not, and there are numerous negative measurement values (around 50%), as seen in Figure 8.1. Also, at any fixed concentration C it is reasonable to assume that the statistical distribution of measurements is approximately normal (Gaussian, bell-shaped-curve) based on both statistical theory considerations and empirical observation. Accordingly, using the normal distribution model with the fitted mean and standard deviation functions to obtain fitted percentiles of the data distributions as functions of C is appropriate; the 5th and 95th percentile functions are shown in Figure 8.2.

8.2.1.1 Critical Level (*L$_C$*)

Currie's first (and lowest) limit is (Critical Level) *L$_C$* (aka Decision Limit). *L$_C$* is fundamentally a way of making a decision regarding when a measured result may be distinguished from a measurement of the same material with no concentration of analyte. It comes from a statistical test of the null hypothesis that the response is not different from the mean response at C=0 (the alternative to the null hypothesis being that it is different).

If an instrument measuring beryllium is calibrated with standards prepared using dissolved sampling media or the equivalent (in our case GhostWipes™, Environmental Express, Mt Pleasant, SC), this is the same as the null hypothesis that the response is not statistically significantly different from that of a media blank (a routine type of QC sample). Such preparation of calibration standards using dissolved sampling media is not unheard of. It is not standard practice, though, particularly for laboratories handling samples from a variety of sources, as a different calibration would be required for each medium. This points out another practical limitation of applying Currie's Concepts to the laboratory practice of producing "detection limit" values, and in turn the potential misuse of non-media-specific limits in decisions using media in which the low-concentration statistical properties vary.

With any such statistical test there is an error rate, called the significance level or the False Positive Rate (FPR, α). This is, on average, the percentage of measurements from samples with true concentrations of zero that exceed L_C. Selection of the false-positive error rate is the sole criterion for L_C; the computation does, of course, depend on the number of data values. In his examples Currie employed $\alpha=0.05$ (a 5% FPR). This is equivalent to computing L_C as a 95% upper prediction limit (UPL) based on the fitted distribution of data at $C=0$, and declaring a measurement to be a "detect" if the instrument response exceeds that upper prediction limit. In Figure 8.3, the 95% upper prediction

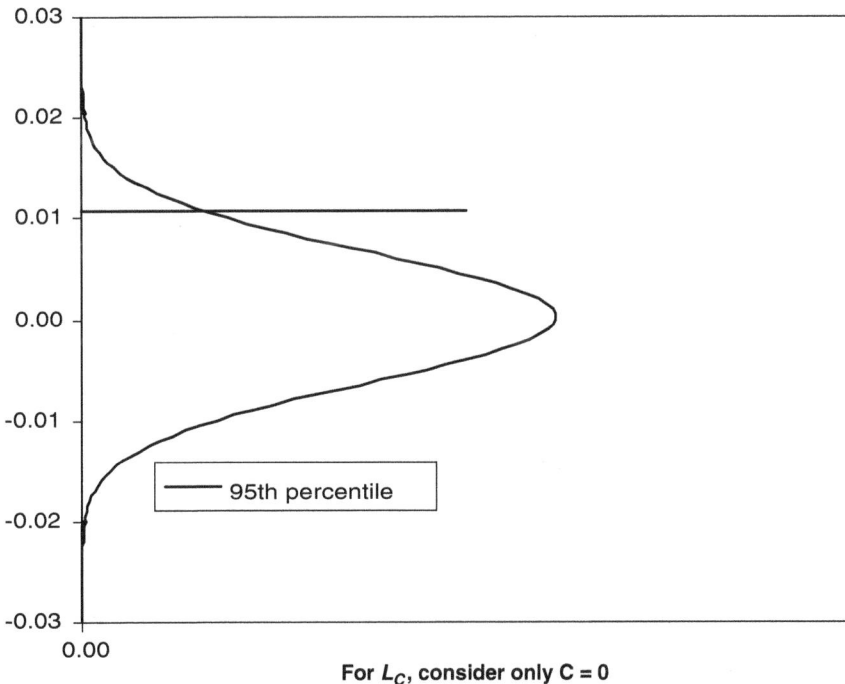

Figure 8.3 Idealized L_C computation for $\alpha=5\%$, $L_C=0.0108$.

limit (5% FPR) has been used, giving $L_C = 0.0108$ µg. In this idealized example, we ignore the difference between the 95th percentile of the fitted distribution and the 95% UPL. With a large dataset such as this one, the difference between these is negligible; with smaller datasets, such as those usually used for laboratory "limit" determinations (*i.e.* $n = 7$), this difference and technical complication in calculation cannot (should not) be ignored.

A 5% FPR, however, should almost never be found acceptable in practice. The reason is that a data user (and the many data users of a particular laboratory and its limit value) is not using the limit value only once or making only one single-datum decision, but will typically be making hundreds or thousands of decisions using the limit, and 5% false positives is simply too many. Accordingly, where multiple detection decisions using one L_C are being made and these detection decisions have significant consequences, a lower FPR should generally be utilized; 1% is fairly standard, perhaps reflecting the Type I error rate requirement for individual comparisons in US EPA groundwater monitoring regulation (see, for example, 40 CFR Part 264.97(i)(2)). There is precedent in that regulation and the associated guidance for taking the total number of decisions being made into account in setting the individual test significance level.[6] Conversely, there can be projects and decisions where a 5% (or perhaps an even higher) FPR may be appropriate, in order to decrease the false negative rate. Section 8.3.1.2 includes a further discussion of the trade-offs between false positive and false negative rates in the context in which decisions are made from individual measurements.

The role of L_C inside the laboratory is in describing (in short-hand) the point at which the FPR reaches the selected rate and what that acceptable rate is. Results below L_C are considered by the laboratory to be unreliable in identifying samples containing beryllium. An L_C type value (commonly either L_C or one-half the L_C) is often used in laboratories to censor data prior to uploading from instrument systems to its Laboratory Information Management System (LIMS). This internal data censoring then becomes a limitation on the laboratory's ability to provide fully uncensored data to data users.

Where L_C is used to censor data (or any other RL for that matter), the consequence is having a complementary False Negative Rate (FNR, β), which results in the need for the second of Currie's Concepts.

8.2.1.2   Detection Level (L_D)

Currie's L_D (*a priori* Detection Limit) is defined as the lowest level at which the percentage of data censored at L_C meets the specified FNR criterion. In statistical terminology the *power* of a statistical test is $(1 - \beta)$, which is the probability of correctly rejecting the false null hypothesis and deciding that indeed there is beryllium present. The FNR depends on the actual concentration; at L_C the approximate FNR is 50% (assuming a normal distribution and no bias, half of the measured results would be below the L_C and thus censored). Moving up from L_C the FNR goes down, until eventually an amount of beryllium is reached at which enough of the distribution of measured results is above L_C to meet the FNR criterion. In other language, L_D is the concentration of an analyte that can *reliably* be detected, or that can be *detected at specified confidence level* $(1 - \beta)$.

The logic is this: having set L_C and decided to censor measured results below it, how much beryllium must be in the sample tested to reliably produce measured values above L_C? In our example (see Figure 8.4), we use $\beta=5\%$ (as did Currie in his original work); L_D is determined to be 0.0217 µg (again using the hybrid model for standard deviation and a 5% FPR).

Note that, in theory, L_C is the critical value on the vertical response axis (*e.g.* peak height or peak area), whereas L_D is on the horizontal concentration axis (*e.g.* as-made, true, actual). In the example, this measurement system makes it impractical to deal with "raw" instrument response (peak height) due to shifting raw response, so instrument response is converted through instrument calibration to concentration. This conversion of raw data prior to calculation of L_C is common and one of the practical compromises from Currie's theories. In this dataset the mean response line is very nearly (Measured=C) perfect (no bias) and therefore this distinction is not important. However, this is not always the case, and it is therefore important to understand this distinction (and nuance in the Currie Concepts). L_C is appropriately an as-measured amount, while L_D is intended to represent a "true" (as-made) concentration. Actual practice, in terms

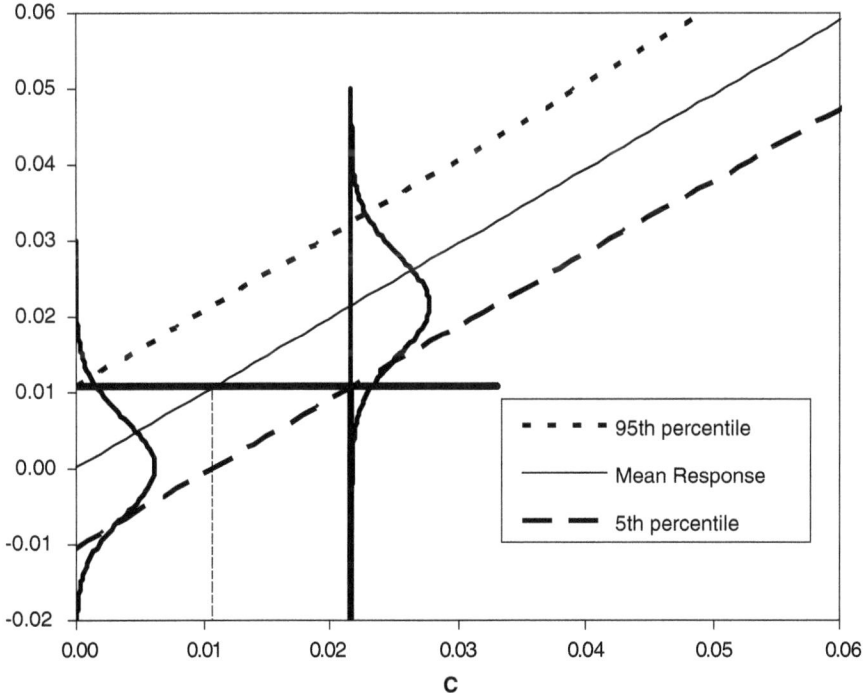

Figure 8.4 Idealized L_C computation for $\alpha = 5\%$, $L_C = 0.0108$.

of how these and their derivative terms are represented, varies widely. Misuse is most detrimental where there is uncorrected bias in the test method.

If data are not censored, then false negatives are not an issue and even negative results may be reported; no criteria for FPR or FNR are required. Provision of uncensored data reported along with the L_C value (and L_D) is a middle ground allowing for full laboratory communication of data reliability and single result decisions, as well as not interfering with large dataset decisions. It should also be noted that Currie did not intend L_D to be used for making data censoring decisions. Rather, one intended use was in study planning. For example, if an investigator needs to be able to "reliably" detect beryllium at concentrations of 0.01 μg (with FPR = 5% and FNR = 5%), the example method/instrument system/laboratory would not be appropriate as it did not reach this level of reliability until 0.0217 μg, and that would require using L_C = 0.0108 μg as the decision level or censoring point. In general, when making a false-negative error is problematic, L_D becomes important.

As noted previously, labs may use L_C or something related for internal (LIMS) censoring, but the censoring in analytical reports may be completely unrelated to the statistical properties of the measurement process or the reported results. This is discussed further in the context of "Reporting Limits" (Section 8.2.4) and Measurement Quality Objectives (Section 8.3.1.2).

In order to communicate data detection limitations accurately and completely, a laboratory should provide the user with two limit values (L_C and L_D), especially if it censors data. In supplying this second limit value in a report on testing, the laboratory adds to its communication on data reliability: the lowest real concentration at which the percentage of measured results below L_C (and censored) reaches the FNR criterion and the FNR the laboratory (or client, if project-specific) finds to be acceptable. Where a laboratory censors data at an RL and does not provide these limit values, the data user will be either uncertain of the false negative and false positive rates at any concentration, or perhaps unaware that there are false negative and false positive risks at all. Having stated this, we remind the reader that, typically, L_C and L_D are determined from off-line analyses of spiked reagent water samples; as a consequence, their statistical properties (and the reliability they are intended to represent) are generally different from that of actual media samples containing matrix interferences, and therefore data users must use caution where utilizing laboratory limit values to represent overall (*i.e.* sampling, sample handling, sample preparation and sample testing) reliability.

It should be noted that reporting an L_D where data have been censored is not as common as might be expected. The primary reason an L_D type limit value is not more frequently included in analytical reports is that many data users, regulatory authorities, and laboratory accrediting agencies are not so concerned about false negative testing results (or are willing to accept the consequences of high false negative rates on decisions) and therefore neither are the laboratories. Laboratories with a higher degree of sophistication, or those supplying data to data users where false negative errors are critical, are more likely to develop and provide these values. Where no L_D has been supplied, data users often double L_C as an estimate. This has a somewhat sound basis in theory, although bias, non-constant standard deviation and other potential errors limit the validity of this estimate.

8.2.1.3 Summary of Currie's Detection Concepts

To summarize, Currie's two concepts related to null hypothesis testing and detection limits are as follows:

- The Critical Level sets the measured level where result values can (based on a selected FPR) be differentiated from results that might have come from testing a sample that contained no beryllium
- The Detection Level sets the lowest actual amount the system can reliably determine to be non-zero or different from blanks (based on a selected FNR) when data are censored at the Critical Level

Censoring result data results in false negatives, which results in the need for a Detection Level (and its provision to data users). These properties are lost, of course, when laboratory performance is summarized by a single RL, the properties of which must be established by the data user.

In moving from Currie's statistically-based concepts of detection to laboratory calculated limit values actually used in analytical reports, much subtle, though important, statistical rigor may be lost. Already noted are the failure to match the matrix of the calibration and limit value to that of the reported result, failure to identify and account for a non-constant relationship of response standard deviation to concentration (using a multi-concentration evaluation), and failure to address statistical uncertainties in estimation of the various quantities involved, as well simplification as regards as-measured Critical Levels *vs.* as-made Detection Levels. In practice, very simplified Critical Level-type limit values are most often developed by regulatory authorities and accrediting agencies and commonly used by laboratories. The Detection Level is less commonly calculated or reported by laboratories, but is an unavoidable consequence of the application of the Critical Level. Both limit types are important concepts for data users to understand, but are aligned with making decisions based on – and communicating information from the laboratory about – single analytical result values.

Data producers and data users need to: understand the Currie Concepts (null hypothesis testing, false positives, false negatives, and the limit values that describe them); understand how a laboratory develops any limit values it uses on reports; and understand the FPR and FNR (if important to the decision) associated with the provided limits. They then need to critically weigh their appropriateness in, and effects on, real-world decisions.

8.2.1.4 Other Issues and Subtleties

As noted in the discussion above on Currie Concepts, the example dataset and calculations used are idealized and simplified, though the limit values determined are not unreasonable for beryllium by ICP-AES. Using 100 independent samples and analyses obtained from a single instrument over a short period of time is also not typical relative to how laboratories develop such limits. The dataset ($n = 7$ is typical) and the study designs used by laboratories are nearly always more limited, and are typically performed at a single concentration (making an assumption of constant standard deviation). The example's study samples, made from spiked wipe media, produce matrix-matched limit values; laboratories testing samples from a number of different media will rarely media-match their limit studies. Spiking with completely dissolved beryllium in nitric acid solution, as was done in the example's study, excludes any variability related to the chemical speciation of beryllium, which may be a significant factor in real-world samples, but of course laboratories will also use dissolved beryllium in their usual limit studies. This study also avoids any extra variation due to matrix interferences from other metals or substances, which can and do affect the response even at C=0. Likewise, laboratories do not typically add interferences to limit study samples. A study taking these additional factors into account would surely tend to increase all the values (L_C, L_D, and L_Q) due to increased variability in the data.

An additional statistical requirement, not addressed in the example, is that of accounting for statistical uncertainty in the mean and standard deviation. Handling the statistical uncertainty in the mean response function is straightforward, but estimating and incorporating the relationship between standard deviation and concentration presents technical challenges. Most procedures in use today, including the procedure most used in the US (the USEPA MDL; see Section 8.2.2), do not account for either of these uncertainties adequately.

It must also be noted that typical limit studies are non-blind. The analyst knows that the sample is for purposes of the study and what the expected concentration is. This may improve the laboratory's apparent performance, especially through elimination of gross errors, as compared to routine sample testing, but this fact should make a data user cautious about accepting a laboratory's stated limit values as being representative of its routine performance with real-world samples. Using double-blind PE samples at concentrations approximating the limit values provided by laboratories (with the data being reported without censoring) is one way that projects may evaluate the relevance of a laboratory's limit values and estimate the magnitude of variables not included in the laboratory's report, and may allow the data user to develop alternative and perhaps more realistic limits, particularly where specific media and/or interference effects are anticipated.[7]

Finally, limit value study designs in laboratories do not usually include allowance for longer term variation, and may or may not include between-instrument variability, let alone between-laboratory variation. Each of these may be important for project purposes. In particular, L_D is a concept useful for study planning; it should therefore be determined in a fashion that takes into account these and other sources of variation. ASTM International has been a leader in requiring inter-laboratory studies for all test methods and is an excellent source of information that may be used to compare the qualities of alternative methods.

8.2.1.5 Currie's Quantifiable Level (L_Q)

Currie's third Concept is that of the Quantifiable Level (L_Q) (aka Determination Limit). This is the true concentration above which the relative standard deviation (RSD) of the distribution of measured values is less than a specified value. Currie suggested 10% RSD as a commonly used value of that criterion in 1968, and that 10% is still cited frequently. Figure 8.5 illustrates this concept. The inset reproduces the mean response and standard deviation functions shown in Figure 8.1, and adds the RSD, which is standard deviation divided by mean. The larger plot repeats that RSD curve, identifying the concentration C at which RSD crosses below 10%. That value is $L_Q = 0.0779$ μg in our example.

With these data, the fitted standard deviation function is $SD^2 = 0.0065^2 + (0.0531C)^2$; at C=0 the standard deviation is 0.0065 μg, whereas for large concentrations, the RSD approaches 5.31%. Note that as C approaches 0, the standard deviation decreases to a positive value, but the RSD approaches infinity. This results from the fact that the standard deviation at C=0 is positive, whereas the denominator in the RSD calculation becomes close to zero. This is sometimes mistakenly taken to mean that measurements become very imprecise at very low concentrations. It is true that measurements at low concentrations are very imprecise when considering precision in *relative* terms,

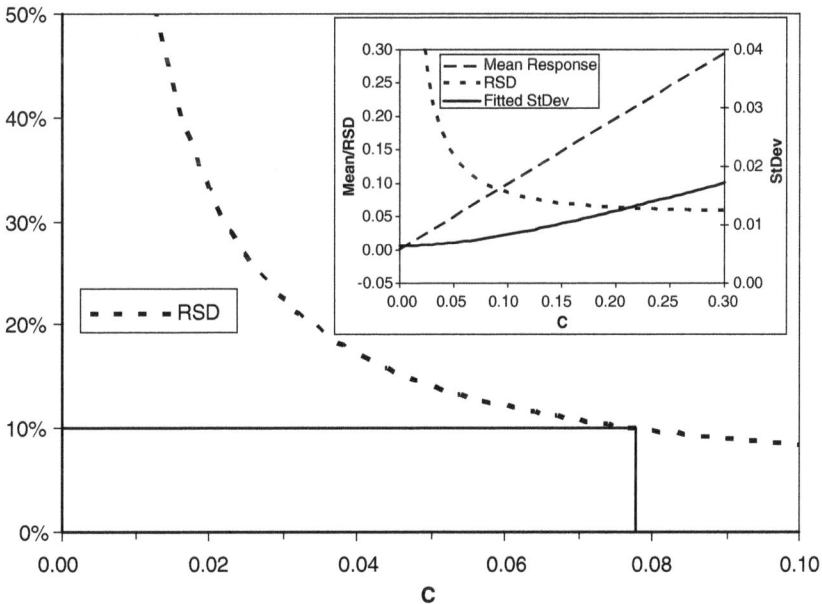

Figure 8.5 Idealized L_Q computation for 10% RSD, $L_Q = 0.0779$.

but in *absolute* terms, the standard deviation most often becomes smaller as the concentration approaches zero, and thus the measurements become more precise in absolute terms.

It is conceivable that 10% RSD may never be achieved for any concentration in some beryllium testing matrices/methods/laboratories, though many study cases do demonstrate this level of precision. In such situations, the L_Q criterion is sometimes relaxed to 15%, 20%, or even beyond. A normal distribution has non-negligible probabilities out to somewhat beyond ±3 standard deviations, which should discourage use of an L_Q criterion higher than 30% RSD. Conversely, measurements could conceivably reach 5% or 1% RSD at somewhat higher concentrations than where 10% RSD is first reached. The project objectives for precision should be considered in relationship to all project criteria that need to be balanced in method selection: interferences, ruggedness, bias, FPR, FNR, cost, availability, *etc.*

It is important to note at this point that Currie's Concept for quantitation specifically addresses precision, but not bias (which is not a significant concern in radiological testing). In common use, however, quantitation is often viewed as incorporating control of accuracy, which includes both precision and bias. In other words, quantitation reliability is conceptually how closely the reported value must be to the "actual" value. Bias is discussed in Section 8.3.1.1.

8.2.2 Implementations of Currie's Concepts: the US EPA MDL

Two years after Currie's paper appeared, Hubaux and Vos[8] applied the L_C and L_D concepts to chemical analyses incorporating linear calibration, explicitly accounting for the uncertainty in both slope and intercept of the linear calibration line, though assuming constant standard deviation. This results in improved rigor in the transition from Currie's theory to analytical practice, with only modestly more complicated computations than those suggested in our Figures 8.3 and 8.4. Clayton and colleagues[9] reported and fixed a technical flaw in the Hubaux–Vos development; see the detailed history and technical discussion provided by Gibbons and Coleman.[10] Both these advances are developed in the context of multi-concentration calibration designs. To date, however, the vast majority of regulatory bodies and laboratory accrediting agencies have not adopted procedures that included calibration design, and therefore laboratories also have not.

8.2.2.1 History of the MDL

In 1981, Glaser and co-workers in the US EPA presented the *Method Detection Limit* (MDL).[11] In their paper, the MDL "*is defined as the minimum concentration of a substance that can be identified, measured, and reported with 99% confidence that the analyte concentration is greater than zero*, [much like Currie's L_C] *and is determined from replicate analyses of a sample of a given matrix containing the analyte.*" The paper's subtitle – describing the MDL as "*that concentration of the analyte that can be detected at a specified confidence level"* – more nearly parallels the language Currie used to define the L_D concept. Their procedure is well-known in the US, having been codified by the US EPA Office of Water into the Code of Federal Regulations (40 CFR 136 Appendix B). Subsequently, the MDL has become the nearly universal de facto standard, pre-empting other approaches or improvements.

In that 1981 paper, the authors do start out allowing the response standard deviation to be a general polynomial function of concentration, but eventually reduce the model to an assumed constant standard deviation "for obvious economic reasons." They avoid identifying that constant with the standard deviation of responses at zero, however, explicitly wishing to avoid considering the possibility of negative responses at $C = 0$. Thus MDL uses a small positive concentration as the basis for the study and assumes that the standard deviation at zero will be the same as at the selected concentration. From their development, it appears that the MDL would have the stated L_D-like property of detection with high confidence only so long as $L_C = 0$, which is unreasonable. Therefore, it is reasonable to assume the authors intended to derive an L_C type limit. The examples and tabled values given in the paper are all for organics analyses, most of which have the preliminary qualitative identification step discussed in Section 8.1.2. In such analyses, quantitative data at $C = 0$ are rarely available and $L_C = 0$ is inconceivable. For all these reasons and given the use of MDL in laboratories as a primary censoring limit, it is reasonable to assume that the MDL is a type of Critical Level.

8.2.2.2 Properties of the MDL

Unlike the procedures for calculation of Hubaux–Vos and Clayton *et al.*, the MDL is derived from a single-concentration study design and calculation. The procedure is well-known: a preliminary guess of the MDL is made, seven aliquots of reagent water (or perhaps another matrix) are spiked at that guessed concentration, the standard deviation of the replicates is determined, and the MDL is that standard deviation multiplied by the 99th percentile of a Student t-distribution with six degrees of freedom (*i.e.* 3.14). If the resultant MDL is excessively far from the original guess, the procedure is iterated (a tacit admission that standard deviations in fact can vary with concentration).

The MDL was used as the basis in 1993 for *Minimum Level* (ML), and before that in 1987 for the *Practical Quantitation Limit* (PQL). The ML is taken as 3.18 times the MDL, which makes it ten times the standard deviation used in deriving the ML, thus targeting the concentration where RSD is 10% and apparently adopting the Currie's L_Q precision criterion. Note, however, that there is a clear fallacy involved in the ML. Nominally the ML should be the estimated concentration at which RSD is 10%, but that relies heavily on the assumption of constant standard deviation over a rather wide range (generally a wider range than assumed when using the same standard deviation for L_C and

L_D derivations). Unfortunately, every chemical analyzed by any quantitative method has a finite ML, whether or not the RSD ever in fact decreases to 10%.

Otherwise, all of the issues and subtleties described in Section 8.2.1 apply to the MDL and ML discussions as well. These limitations in rigor are particularly problematic where laboratory MDLs become *de facto* regulatory compliance limits, as occurs where risk-based compliance limits are lower than analytical capabilities. Any detection recorded on the laboratory report may be deemed a violation of the risk-based standard because the value reported is greater than the standard.

An additional reality is that, because of the prominence of place accorded the MDL, a laboratory's ability to "meet" the anticipated MDL is frequently used as a criterion in deciding whether that laboratory will provide data of adequate sensitivity and precision for a particular project, or whether that laboratory can be certified by some agency or body. This has two adverse effects. The first is that MDL determinations are not only made off-line using non-blind analyses, but that they can often be made with an eye on the contract or certification target. The second, which follows from the first, is that a knowledgeable analyst, who realizes that standard deviation varies with spiking concentration, has some control over the estimated MDL through choice of spiking concentration for the study design. In addition to all of the above and perhaps most importantly, one must recognize that standard deviation estimates based on only seven (not wholly independent) observations have a high degree of statistical uncertainty and are not reliable estimates of routine capabilities.

8.2.2.3   Current Status

The statistical and conceptual simplifications, as well as the limited study designs, coupled with expanded application of MDL and ML in regulatory decisions, eventually led to a 1999 lawsuit filed against the EPA by several industry groups (*Alliance of Automobile Manufacturers* v. *EPA, DC Cir., No. 99–1420*) based on EPA's revisions of a test used to measure mercury concentrations at low levels and challenging the scientific legitimacy of the MDL. This resulted in a Settlement Agreement that required EPA to reassess and potentially revise the procedures used to determine detection and quantitation limits for use in EPA's Clean Water Act programs by 1 November 2004.

The first response by EPA to this court-ordered review was a minor revision of 40 CFR 136 Appendix B set forth as a proposed rule-making in March 2003.[12] This, and its supporting technical document, received extensive criticism and were withdrawn the following year. Subsequently (December 2004), a Federal Advisory Committee on Detection and Quantitation Approaches and Uses in Clean Water Act Programs (FACDQ) was convened to provide advice and recommendations to the Office of Water. The Final Report of the FACDQ[13] was submitted to EPA in December, 2007. This was presented as a set of consensus resolutions, which required all FACDQ members, including

EPA, to agree or abstain. Notable agreements within the FACDQ include the following:

(1) A FPR not greater than 1% should be used for the detection limit ("DL") by US EPA Office of Water.
(2) The quantitation limit ("QL") – to be set as a multiplier of the detection limit – should be required to assure a FNR not greater than 5% as well as address precision concerns similar to those of Currie's L_Q.

The Final Report also discusses many issues on which the FACDQ did not achieve consensus. The FACDQ did not reach agreement that the single-laboratory detection limit procedure it developed should be adopted by the Office of Water. However, it did agree, by consensus, that if or when data are reported below QL, then the data points that fall between DL and QL would be reported, for example, as Detected but Not Quantified (DNQ). It also recommended that *"EPA act to develop an alternative to the current 40 CFR Part 136 Appendix B procedure"*.

The Final Report clearly mixes Currie's Concepts and associated terms. It starts out by adopting definitions of "Detection Limit (DL)" that correspond to L_C rather than L_D in Currie's system; its "Quantitation Limit (QL)" definition reasonably corresponds to Currie's L_Q. However, it explicitly avoids L_D: *"the Committee agreed that while the concept of the L_D was important, it would be acceptable not to derive an L_D, on the condition that the false negative error rate at the Detection Limit was acceptable for results at the Quantitation Limit"* [in Section 1.5].[13] Recall, though, that its definition of DL is phrased in terms equivalent to FPRs, not FNRs; later the report makes it clear that the target FNR should be attained at the QL, with detection decisions being made using the DL. Regarding FNRs, *"the Committee agreed to ignore false negatives for detection but instead included them in the concept of quantitation as a condition of dropping L_D."* Measured values less than DL should be reported as "less-thans" or U or ND; measured values between DL and QL should be reported as "J" values or as DNQ. Since receiving the recommendations of the FACDQ, the Office of Water has agreed to conduct a pilot study to further evaluate the procedure developed by the FACDQ or alternative procedures. It is expected that this study will be completed in 2009 and that EPA may propose a revision to the MDL procedure in Part 136 as early as 2010.

8.2.3 Recent Advances: ASTM Contributions

A good deal of sophistication in implementation of Currie's Concepts has developed since 1990, despite the litigation of USEPA MDL. Advances as described in Section 8.2.2 and the derivation by Rocke and Lorenzato[4] of a conceptually sound and empirically supported (hybrid) model for the relationship between standard deviation and true concentration have contributed. Gibbons and Coleman[10] detail several additional advances, such as modifying

L_C to accommodate multiple future detection decisions, allowing for non-constant standard deviations in both "limit" and calibration determinations, and taking between-laboratory variation into account in derivations paralleling Currie's L_C, L_D, and L_Q, and others. These advances have been incorporated into two ASTM International standards – the *Interlaboratory Detection Estimate* (IDE)[5] and *Interlaboratory Quantitation Estimate* (IQE).[14] The ASTM Committee (D-19) that developed the IDE and IQE procedures has also developed and made available software (DQCALC[15]) for performing the calculation of IDE and IQE. Based on generally available spreadsheet procedures, this user-friendly software automates the complications incurred in rigorous implementation of Currie's Concepts. This software, as is, may also be used to calculate single laboratory L_C, L_D, and L_Q; the Committee is currently developing the associated technical standards for Within-Laboratory Detection and Quantitation Estimates (WDE and WQE).

8.2.4 "Reporting Limits"

Preceding subsections of Section 8.2 have dealt with the history of decision, detection, and quantification levels, starting with Currie's Concepts, proceeding through their incomplete implementations, the US EPA's MDL and its dominant position in the US, and the subsequent legal challenge to the MDL due to its applied use in regulatory compliance, ending with the more recent and rigorous ASTM products being made available as viable alternatives. Somewhat related to these concepts is the idea of a *Reporting Limit* (RL), as discussed in Section 8.1.1.

A general definition of this commonly used term is the following: the RL is a limit value established by contract or agreement between a laboratory and the data user below which data are censored. In the absence of an agreement, a laboratory will use its customary (routine) RL, as established by its written quality system. Laboratories determine their routine RLs based on a variety of factors including instrument sensitivity, calibration range, sample amount (volume), and dilutions. Some laboratories base their RL on some multiple of the MDL, which can lead to fluctuation when the MDL is recalculated. Others set the RL somewhat higher to avoid that fluctuation.

Values less than the RL are reported as " $<xx$ " where xx is the numerical value of the RL. The RL can vary among analyses in the same run when dilutions are involved. For this reason and for the convenience of the data user, the " $<xx$ " reporting format is to be encouraged, as opposed to just stating "U" or "ND" without specifying the numerical RL.

Laboratory accreditation generally requires that the laboratory have a written policy and procedure for establishing its routine RL. Typically such a procedure will utilize detection and/or quantitation limits in establishing these RL values, although they may be based on other criteria altogether (*e.g.* a value specified in the method of testing, a risk-based regulatory limit, or the concentration of the lowest calibration standard used in the testing). The data user, where the laboratory uses its routine RL, should understand how the reporting limit was derived and the qualities, such as analytical variability, that may be expected from data above the RL. Alternately, if the data user should require RLs other than the laboratory's standard RL, the method used by the laboratory to determine that it can provide data of adequate quality above the alternate RL should be determined.

Many project and regulatory programs require that data be reported at least as low as a prescribed concentration. These client-specified RLs often come from a project data quality objective (DQO) process (see Section 8.3, for example), or from particular health or environmental risk calculations. Laboratories will usually accommodate these client requests and provide data to the requested limit, so long as it is no lower than the MDL (Critical Level). If such requests are supplied to the laboratory during the planning process, the laboratory may suggest an alternative analysis technique (*e.g.* ICP-MS instead of ICP-AES), may indicate that a larger sample be taken, or may be able to utilize a larger sample aliquot to accommodate lower-than-routine RL requests. Once testing has been completed, in most cases, laboratories will not accommodate client-requested RLs that are lower than the laboratory's MDL, and in many cases are prohibited by accreditation and certification requirements from doing so. Most laboratories will also qualify or "flag" all data reported below "quantitation" limits as well.

To the extent that the laboratory's RL may provide an indication of the quality (see Section 8.3) of its above-RL data, the data user should inquire about the RL and the procedure used to derive it even if the user will ultimately request uncensored data. For instance, in a beryllium facility investigation aimed at determining whether the 95th percentile of the distribution of beryllium concentrations is less than $0.2\,\mu g$ per sample (wipe), the laboratory should be willing to provide an RL considerably lower than that value. If the laboratory's routine RL is $0.2\,\mu g$ or higher, the data are not likely to be usable.

We should mention that sometimes a project specification for a RL includes precision criteria resembling those of L_Q in addition to (or perhaps in place of) the FPR and FNR criteria addressed by L_C and L_D. When precision criteria are included, the resulting RL is likely to be higher than it would have been without those criteria, but it may be difficult to infer further information about the quality of the data from such an elevated RL.

8.3 Data and Measurement Quality Objectives

8.3.1 Evaluation of Data Quality Objectives

Around 1990, the US EPA introduced the idea of Data (or Decision) Quality Objectives (DQOs). The goal of this approach is that investigators should consider *quantitatively* the nature of the data that might be obtained in a study while the study is still in the *planning stage*. There is a formal seven-step DQO process:

(1) State the problem.
(2) Identify the goal of the study.

(3) Identify information inputs.
(4) Define the boundaries of the study.
(5) Develop the analytical approach.
(6) Specify performance or acceptance criteria.
(7) Optimize the plan for obtaining data.

This should all be done *before* expending resources on data collection.

Measurement Quality Objectives (MQOs) refer to the subset of DQO considerations most directly related to the quantitative aspects of the DQO process, particularly part (7). MQOs are discussed extensively in the MARLAP manual.[16]

In principle, one should do a DQO evaluation for every project. In practice, this requirement would be burdensome, particularly for common and routine tasks. Moreover, with respect to projects incorporating laboratory analyses, busy analytical laboratories may well balk at implementing different sets of requirements for each new project. On the other hand, developing DQOs carefully can aid in clarifying where one may most profitably expend analytical budgets; see in particular discussions of the TRIAD approach to site characterization recently developed and promoted by the EPA's Technology Innovation Office.[17] A key feature of the DQO/MQO approach is that decisions regarding data treatment (including data reporting) and measurement quality requirements are steps in a process that starts with asking what the decision requirements are for a particular project, and are constructed from that point of view.

8.3.1.1   Data Quality Indicators

The term "data quality" appears in the final subsection of Section 8.2. This term embodies a variety of concepts, which comprise the third component of the DQO/MQO system: Data Quality Indicators (DQIs). The primary DQIs are data representativeness and completeness, bias and variation (precision), and data comparability and sensitivity.

Data representativeness refers to taking enough samples at appropriate locations to provide reliable answers to the questions that motivated the data collection effort in the first place. This issue has to do with the design of the data collection effort; this is beyond the scope of this chapter, although data should be reported in such a manner as to allow a reviewer to evaluate the representativeness of the sampling plan utilized.

Data completeness refers to the proportion of samples anticipated under the sampling plan that are actually collected and analyzed, producing data that are usable for the desired purpose. In planning, consideration should be given to the likelihood of failure in sampling and analysis, and plan accordingly to assure that an adequate number of samples will be available for the required reliability of decisions. Again, one should be able to evaluate completeness from appropriate data and/or project reports. In particular, if samples were not obtained at specified locations, or samples were obtained but could not be analyzed successfully, one should investigate the circumstances.

The other DQIs are more closely related to the analytical issues discussed in Section 8.2. These include evaluations of the bias and precision of an analytical method. These metrics can provide an alternate way of looking at the statistical properties of data that one might anticipate receiving from a particular laboratory.

Bias refers to the difference between the average value of a measurement obtained at a concentration C and C itself. This is of particular interest, and can be problematic, in beryllium studies, in that one's view of the "true concentration" of beryllium contained in a sample can vary according to the particular chemical speciation involved and the sample preparation used, as is discussed in other chapters. From a statistical standpoint alone, one is tempted to "punt" on this issue, and simply suggest that in general the "true" value for a particular method (including digestion) might reasonably be considered to be the mean value of measurements obtained using that method. In other words, one might automatically set the bias to zero by definition, relegating all considerations of what is the "true" concentration for a particular analysis to the selection of digestion method and other sample preparation issues for the analysis. These issues are discussed in greater detail in Chapter 4. In beryllium testing, bias may be incorporated through the sampling process, the preparation process, and the analytical process. Described in other chapters, these biases are generally not "corrected for" in reported data (or associated data qualifiers such as limit values). Bias in the design of a sampling plan cannot be evaluated by evaluating only laboratory reports or data packages.

Precision refers to the variability of measurements. This can include several components (variance components in statistical terminology): variation within a given instrument in a given laboratory on a given day; between-instrument variation in that same laboratory; longer-term variation in that laboratory, variation among laboratories; variation due to differences in sample matrices; and so on. As discussed previously not all of these are evaluated in typical "limit" studies, or in routine quality control or studies designed to assess precision (and bias) over the working range of the method of testing. In principle, one ought to take these various forms of variation into account in evaluating the overall precision of analyses. Many of these can be accommodated through the design of blind or double-blind PE studies, or project-incorporated field duplicate quality control and field blank quality control samples.

The related concept *accuracy* is sometimes included as a DQI as well, integrating precision and bias. A typical mathematic expression of accuracy is in terms of Mean Squared Error (MSE), which is the square root of the average squared deviation of the measured value from the true concentration; as an equation, $MSE^2 = (StDev)^2 + (Bias)^2$. This is the same as the standard deviation if the bias is zero, perhaps by definition as suggested above. On the other hand, the equation suggests that a system with some bias but less variation may be preferable to one with less or no bias but greater variation. This concept may apply, for instance, when attempting to "correct" analytical results for matrix interference, where the correction does indeed make the average value of measurements more nearly what they ought to be but increases the variability considerably.

Data comparability refers to the idea that data from different laboratories should be comparable, as should data from one laboratory obtained at different times. On an inter-laboratory level, this has been evaluated through inter-laboratory studies conducted by various bodies. These studies are costly and difficult to coordinate, are not frequently repeated, and may become dated as systems of measurement and available technologies improve. This should be taken into consideration when considering the comparability of methods. The comparability of routine quality control data generated by different methods and/or different laboratories can also be used to support or contradict an assumption of comparability, as can other measures such as performance testing results, MDL studies, *etc.*

Sensitivity is closely related to the L_C and L_D concepts, which relate directly to precision at very low concentrations (as discussed at length in Section 8.2). At what concentration can the laboratory be counted on to reliably detect non-zero concentrations low enough for the purposes of the investigation at hand (given a particular sample type, size and method)? Increased sensitivity tends to come at the cost of method ruggedness and management of interferences, and – as with the other DQIs – must be balanced with other DQIs, ruggedness, availability, cost, and throughput capacity.

An open question at this time is how one might best evaluate the data qualities of prospective laboratories and/or methods contending for project use. As discussed in the MARLAP,[16] other information – often utilized by accrediting agencies and generally available – may be of value. However, expertise in evaluation of this information is required. Alternatively, perfor-mance in pre-qualification testing (in project matrices and with prospective methods) provides valuable insight. As indicated previously, ASTM Interna-tional has required for many years that methods of testing be assessed through inter-laboratory studies and the outcomes included in each approved method.

8.3.1.2 DQIs for Single-Datum Decisions

Most of the MQO issues embodied in the DQIs take a relatively simple form when decisions are made using individual measurements, at least from a data reporting point of view. Data representativeness refers to whether the samples are taken appropriately at the right location at the right time or over the right period. Data completeness is a simple question of whether the sampling device worked as expected and the laboratory was able to analyze the sample. Data comparability can be addressed by occasional side-by-side duplicate samples using different sampling devices, possibly sending the samples to different laboratories for comparisons. Evaluating bias for beryllium measurements has already been discussed.

Regarding precision and sensitivity, typically when decisions are made from individual beryllium measurements, the question is whether the concentration (exposure, *etc.*) is above a standard or regulatory criterion (RC). There are two types of errors involved: (A) deciding that the concentration is lower than the RC when it is above; or (B) deciding that the concentration is above the standard when it is below. In carrying out an MQO analysis, one might select an acceptable false decision probability for error A when the concentration is actually at the RC, and also a lower level and acceptable false decision probability for error B when the concentration is actually at a routinely achievable lower level – denoted "RAC" for this discussion. This setup closely resembles the situation with L_D and L_C. Carrying it out is straightforward, and will likely involve an intermediate "action level" ("AL"), lower than the RC, at which one will take some sort of corrective action, without declaring that the individual has been exposed above the standard. Note that the AL should not be the RC itself; if the AL is set equal to the RC, error (A) will occur with around 50% probability when the actual concentration equals the RC. This would be an error comparable with using L_D as a reporting limit.

Where engineering and administrative controls are adequate, exposures are expected to be below the RAC described in the preceding paragraph. Whether one can find an appropriate AL that will balance the probabilities of errors A and B acceptably will depend on the *precision* of measurements made when the concentrations are equal to the RC and the RAC. In carrying out an MQO analysis of this decision setup, one would want to be able to believe that the precisions at both levels are acceptably low. If the laboratory's MDL is low enough to suggest that the standard deviations at both the RAC and the RC ought to be acceptable, the task is done (assuming that one is willing to believe that the MDL study data are reasonably representative of actual project data and routine operations in the laboratory). Evaluation of the laboratory's MDL study design and results, review of ongoing laboratory method blank and precision information, or other checks on the validity of the MDL should be considered; if appropriate, the development of a double-blind PE program may be in order.

Figure 8.6 shows a hypothetical situation in which one would have wide latitude in setting an AL. As discussed previously, it is often reasonable to assume that measurements are distributed normally at fixed concentrations, although the standard deviations in general depend on the concentration. The horizontal axis serves as both the true concentration and measurement axes; negligible analytical bias is assumed. In this example, we set the allowable probability of error (A) (the FNR) equal to 1% and that of error (B) (the FPR) equal to 5%. The distributions of measurements at concentrations equal to the RC (solid line) and the RAC (dashed line) barely overlap; any AL between AL_{min} and AL_{max} will succeed in meeting those criteria.

In this first example, one can easily imagine that the RL would be set equal to the AL chosen. Since any AL between AL_{min} and AL_{max} will satisfy the MQOs, the decision becomes primarily a public or worker relations concern: should AL be set lower, conveying the message of greater worker protection with respect to the published RC at the expense of a slightly higher FPR, or would a higher AL (but still less than the RC) with its lower FPR be wiser? Either way, note that the decision regarding RL has little, if anything, to do with L_C, L_D, or L_Q.

Figure 8.7 is not so comfortable; the precisions of the data distributions are not as low as in Figure 8.6, so that it is impossible to meet both the FNR and FPR criteria. This situation might arise with analytical methods inappropriate for the RC at issue, but it might also arise when attempting to extrapolate from a single sample of limited extent (*e.g.* time duration). In Figure 8.7, if the AL

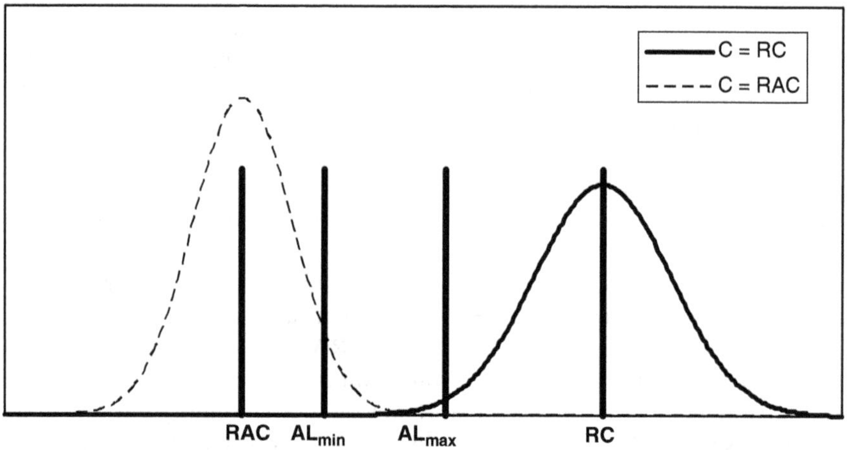

Figure 8.6 MQOs – first example. RC = Regulatory Criterion; RAC = Routinely Achievable Concentration; AL = Action Level.

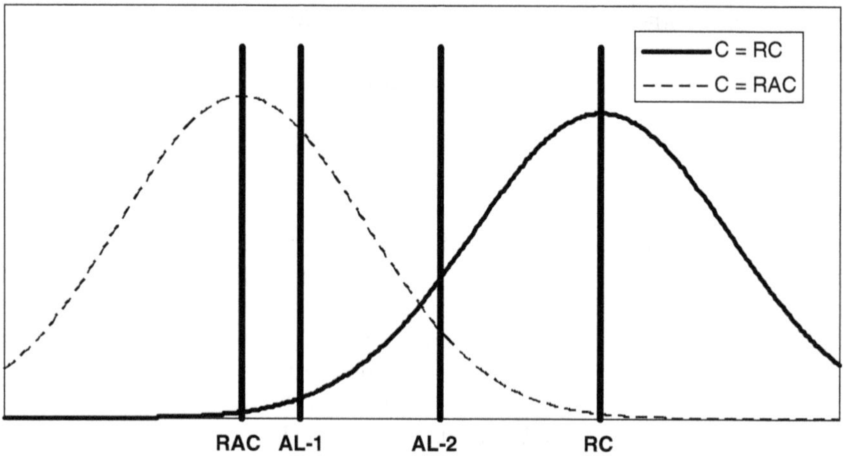

Figure 8.7 MQOs – second example. RC = Regulatory Criterion; RAC = Routinely Achievable Concentration; AL = Action Level.

is set at AL-1 so that the 1% FNR target is met, the FPR becomes 31%. Conversely, if AL is set at AL-2 in order to meet the 5% FPR target, the FNR becomes 11%. If this measurement system is used, compromises must be made with respect to the desired FPR and FNR targets.

8.3.1.3 Evaluation of Datasets

The situation is not nearly so simple when decisions are to be made statistically using entire datasets. In this situation, there is typically a large location-to-location (individual-to-individual, *etc.*) component of variation involved. Whereas data distributions at given, fixed concentrations may reasonably be assumed to be normal, the same is not true of location-to-location variation. Rather, such variation is often assumed to be lognormal in distribution, *i.e.* the logarithms of the concentrations have normal distributions. This leads to a mixed distribution model that has not been studied in nearly the depth that pure normal or pure lognormal models have been studied (see Section 8.4.1).

Log-normal distributions can be notoriously heavy-tailed (right-skewed), with small proportions of rather extreme values. This description fits the sort of data one obtains in facility surveys. On the other hand, there are other families of perhaps more nicely behaved skewed distributions that might be considered as well. It is difficult to select among the various possibilities; theoretical considerations favor the log-normal family slightly, but empirical evidence is inconclusive.

When data are censored, one may not recognize the mixed distribution aspect of the situation. One may then assume that the overall distribution is log-normal, and use sophisticated log-normal censored data statistical tools in making inferences regarding whether distribution means or 95th percentiles exceed a relevant standard or regulatory criterion. In doing so, one is implicitly assuming that the normal distribution component of analytical variation is negligible compared with the (assumed) log-normal spatial or person-to-person component. This implicit assumption may be reasonable so long as the analytical component of variation is truly dominated by the spatial or person-to-person component. However, recent research suggests that this assumption may be overly conservative.[18] Future research in this area is warranted, including research into appropriate ways of setting and verifying the attainment of appropriate MQOs in this setting.

Alternatively, one may opt for receiving and using uncensored data as described in Section 8.4. Again, setting appropriate MQOs is an open research question in this setting.

8.3.2 Alternatives to "Detection Limits"

One occasionally hears calls for adopting alternatives to the current system of limits. For example, in "Part 26 – Detection Limits: Editorial Comments and

Introduction" of their column *Statistics in Analytical Chemistry* Coleman and Vanatta open by stating bluntly:

"We believe that detection limits (DLs) should go away. . . . Instead, we encourage the scientific community to adopt a policy of reporting every measurement result . . . with a statistically sound estimate of its uncertainty. . . . Then the customer would truly have the data needed to make sound judgment calls. However, people typically shy away from making decisions for which they could be held accountable, perhaps helping to explain the perpetuation of the status quo".[19]

Adopting such a proposal would be essentially equivalent to replacing every *number* reported by a laboratory with an *interval*. Such reporting is common with radiological measurements, where one typically receives (result±error); often the stated error is twice the analytical standard deviation, so that the interval approximates a 95% confidence interval.

In such a system, however, the "error" reported, as with the limit values, depends on the study design, qualities, and quantities of the data used in development of the interval, and the selected interval itself. It is likely to include only analytical variation, perhaps omitting even the error due to sample preparation and most likely omitting any long-term or between-laboratory variation and so on. If history repeats itself, it is also likely that single concentration data will be used (*e.g.* the mid-range QC reagent spike), with the standard deviation assumed constant. Hence the data user would have to be become knowledgeable and aware of subtleties as discussed previously with limit values in order to interpret the information appropriately. Further, if the user required a different interval or a different coverage, the underlying information used to construct the interval would have to be supplied. Moreover, if data were reported in this fashion and such data were made widely available, one would have to provide education to data recipients. Such interpretation advice would need to include dealing with situations where: (a) the interval reported straddles a concentration or exposure criterion or standard; and (b) the interval includes the concentration zero or is even entirely negative, as can happen with beryllium measurements.

8.3.3   Total Measurement Uncertainty

In some circles, there is a push toward ensuring that evaluations of measurement variation include all relevant components of uncertainty. This concept has appeared several times in this chapter already, in discussions of the TRIAD[17] and MQO initiatives, for example, in the discussion of employing entire datasets in Section 8.3.1.3, and in the discussion of alternatives to "detection limits" in Section 8.3.2. One should mention in addition the emphasis placed on this idea by the International Standard Organization in its publications.[1,20,21]

8.4 Using Uncensored Data

Where uncensored datasets are available, they may improve the efficiency of decisions made *via* statistical comparisons. The prime example is use of *Upper Tolerance Limits* (UTLs) in evaluation of compliance with a clean up criterion. A 95%–95% UTL is an upper 95% confidence limit for the 95th percentile of a distribution. In facility surveys, it is common to set up the decision to be made as a statistical hypothesis test, the null hypothesis being that the facility is not (adequately) clean; this is equated with the 95th percentile being at least as large as a Regulatory Criterion (RC). The facility management wants to obtain data that will reject this null hypothesis in favor of the alternate, which is that the 95th percentile is less than the RC.[22] One way of implementing this statistical hypothesis test is to obtain data (using a scientifically sound sampling plan) and compare the UTL computed from those data with the RC. If the UTL is less than the RC, the null hypothesis is rejected and the facility declared clean; otherwise, there is not adequate proof that the facility is clean, and more sampling and/or more housekeeping must be performed.

UTLs are of two types: parametric and non-parametric. With no assumptions about data distributions, one can use a non-parametric or distribution-free UTL, in which one simply sorts the data values and uses the appropriate ordered value as the non-parametric UTL (NPUTL). To achieve 95% confidence for estimating the 95th percentile, one needs at least 59 observations. The NPUTL is the largest value for datasets containing from 59 to 92 observations, the second largest for datasets containing from 93 to 123 observations, and so on. When using NPUTLs, one is not concerned about whether the data are censored or not, as long as the censoring point is a value below the RC; see Gilbert[23] or Gibbons and Coleman[10] for a discussion.

The requirement for 59 observations per facility can be burdensome, however, particularly for smaller facilities, as discussed by Davis and Grams.[7] If one wishes to construct UTLs with fewer than 59 observations, one must use a parametric statistical approach. That is, one assumes a particular distribution family (usually normal), possibly after data transformation (log is most commonly used, resulting in an assumed log-normal distribution), uses standard formulas and/or tables for constructing the UTL, and un-transforms the resulting UTL if a data transformation was used.[10,22] When some or all data are censored, however, there are complications. Dealing with these complications is an active area of research in statistical methodology, the details of which are beyond the scope of this chapter.[24] One should be aware, however, that most such research has targeted the problem of estimating means and standard deviations of distributions, and that simply plugging estimated means and standard deviations into the usual UTL formulas does *not* guarantee that the resulting UTL will maintain the desired confidence level. Typical research reports include language such as "*if the lognormal distribution is not excessively skewed and no more than 70% of observations are censored, 'such-and-such' modification of the usual procedure maintains the desired confidence tolerably well.*" [25] Such censored data parametric procedures can be used with much smaller numbers of observations than 59, and may be most useful for particularly clean, small facilities.

With higher proportions of censored data than approximately 70%, however, the censored data parametric procedures are not available, at least with the current state of research. Even if they can be used, there may be a penalty in statistical efficiency, in that the censored data UTL may be higher than what one might obtain if all data were available, as illustrated by Davis and Grams.[7] In each of the above situations, where censored data limits the statistical application available or its effectiveness, the data user should consider use of uncensored data.

There are several additional circumstances in which one might profitably use uncensored data. These all involve decision-making using whole datasets. Not all involve UTLs, however. Moreover, some of them involve the secondary analysis of data already acquired, for which the primary purpose was individual exposure assessment. When the statistical analysis intended does not involve UTLs and/or when a large number of observations is likely to be available, the motivation to request and use uncensored data is less compelling than when decisions will be made using UTLs. This is fortunate, particularly in the case of the secondary use of data, in that one might thereby avoid requesting different report for the same samples when the data are used for the multiple purposes.

8.4.1   Using Uncensored Data: Technical Issues

8.4.1.1   *Negative Values*

The primary issue arising with uncensored beryllium data is the fact that, due to analytical variation such as that shown in Figure 8.1, one routinely obtains negative values for the analyses of blanks and other samples obtained in very clean environments. These negative values do not represent actual concentrations, of course, but do represent legitimate instrument responses. Ignoring or discarding negative values would result in biasing one's statistical estimates upwards, which is undesirable. A trivial implication of this is that a database designed to hold uncensored beryllium data must be able to accommodate negative values.

A non-trivial implication is that statistical methods designed for data that can take only positive values will require modification. This is the case with the log-normal family of distributions, commonly used for environmental data.[22] This model has some justification in statistical theory and empirical observation; distributions in this family range from nearly symmetric to extremely right-skewed, the latter meaning that extremely high observations can occur, but with low probability. Another family suggested for environmental concentration data is the gamma family. This family is not as easy to use as the lognormal family, but is less "heavy-tailed" than the lognormal family and is preferred by some authors.[26,27] Neither family of distributions allows for negative values without the incorporation of a "shift" parameter. Estimating or fitting such a shift parameter is an extension of the usual application of these distribution families; accordingly, additional research is needed to develop reliable ways to estimate these parameters in a fashion that will maintain the desired confidence of the resulting UTLs tolerably well, at least *in situa*tions where the UTL might be close to the RC, and therefore close control of the UTL confidence is particularly desirable.

8.4.1.2 Research Needs

In addition to the research need identified in the preceding paragraphs, the existence of negative values raises additional research questions. A primary one has to do with what the actual distribution of measurements should be. A shifted log-normal or gamma distribution is an approximation; the actual distribution involves the underlying distribution of the true concentration, which is in principle unobservable, to which is added a normal distribution due to analytical variation, with standard deviation varying with true concentration. This is more complicated than a shifted log-normal or even a shifted gamma distribution; at present, the identification of optimal UTL (or other) statistical procedures for such a mixture distribution is an open question.

But this raises an additional, troubling issue for the use of censored data techniques. These are generally developed assuming that the underlying distribution of *measurements* is lognormal. Since this is not the case, the censored data techniques must therefore also be at best approximate. Therefore, an additional research need is to determine how sensitive the performance of these techniques might be to the model misspecification. One would like to identify, for example, rules of thumb relating the performance of these statistical methods to appropriate functions of data precision and pertinent regulatory criteria or standards. Again, these are at present open research questions.

8.4.2 Using Uncensored Data: Non-technical Issues

We emphasize again that the reporting and use of uncensored data are nonstandard, and that when non-censored data are to be provided, it should be at the specific request of a knowledgeable user and as a supplement to the standard report. Two reasons for this emphasis are described below.

8.4.2.1 Public Relations Issues

There are always concerns regarding public and worker relations that might involve exposure to toxic materials, even when those materials are naturally occurring, as is beryllium in many areas. In the case of beryllium, this concern is ameliorated somewhat by the fact that our typical analyses are looking for total beryllium, only at most a small portion of which is likely to be toxic anthropogenic forms (see Chapter 1). Nonetheless, one should take care with data communication when uncensored (or even censored) data are provided to provide clear but technically sound explanations of what the measurements are and how they should be interpreted.

8.4.2.2 *Accreditation Issues*

The primary purpose of accreditation is to provide an independent means of verifying the quality of the data produced by a laboratory, providing a benefit to both the laboratory and its customers. A typical requirement of accrediting bodies is the establishment of a reporting limit as a means of assuring that numerical values below the calibration range or sensitivity of the analytical system are not reported, or are qualified in some fashion. Section 8.2.4 provides a discussion of the concept of reporting limits, and a general definition. The AIHA, which provides the accreditation services used by most US industrial hygiene laboratories, defines reporting limit as "*the lowest concentration of analyte in a sample that can be reported with a defined, reproducible level of certainty.*"[3] AIHA laboratory quality policy requires that measurements below the reporting limit be censored (*i.e.* reported as "<xx", where xx is the reporting limit).

When an accredited laboratory is asked to provide numerical values below the reporting limit, the laboratory data report must include appropriate caveats regarding such data, to avoid the risk of violating accreditation requirements. As an example, the US EPA's Contract Laboratory Program (CLP) defines specific data qualifiers and when to use them.[28] AIHA quality policy simply states that test results not covered under the accreditation must be clearly identified on the test report.[3] Thus, measurement values below the RL may be provided as long as they are clearly identified as described above, and preferably in a supplemental report (not the primary test report) to avoid confusion. It is important for the laboratory to document its policy regarding when and how such data are provided in its written quality system, and to ensure that the policy complies with applicable accreditation requirements.

8.5 Summary

This chapter has focused on three areas: "Detection Limits" and related concepts; Data and Measurement Quality Objectives; and the use of uncensored data. Detection and other related limit types have a foundation in Currie's Concepts of critical, detectable, and quantifiable levels. These are all defined in statistical terms:

- The critical level L_C is intended to control the false positive rate of decisions regarding the presence of beryllium.
- The detectable level L_D (which Currie called the a *priori* Detection Limit) is intended to be a concentration high enough to be detected using L_C with a satisfactorily low false negative rate.
- The quantifiable level L_Q is intended to be a concentration level above which measurements will have a stated precision (usually 10% RSD) or better.

We emphasize again that these are all defined relative to interpreting *individual measurements*, not to decision-making using entire datasets.

In moving from these concepts to the practice of developing limit values, specific approaches must be selected and trade-offs made. Through use of an example, this chapter introduces the Currie Concepts, examines some statistical tools, and discusses a number of implementation issues (selection of false positive and false negative rates, non-constancy of standard deviations, *etc.*) and technical challenges (*e.g.* incorporating statistical uncertainty in estimated mean response or calibration and standard deviation functions).

The history of limit development, in particular the US EPA's MDL, is discussed at length because of its preeminent and consequently preemptive status, particularly in the technical limitations of the version of the MDL currently (2008) embodied in US Federal Regulation and of how limits derived from it are presented; these are the basis of a lawsuit, which may eventually lead to revisions.

Other, more sophisticated technical developments are reviewed, particularly those spear-headed by ASTM International that allow development of more rigorous limit values where this is important to data use. In addition, alternatives to the current system (or rather horde) of limits, such as interval-based reporting of all analytical results, are described in overview.

The second major area discussed is that of Data and Measurement Quality Objectives (DQOs and MQOs). These involve determining, during the planning stages of an investigation, what measurement statistical properties will be needed of the laboratory's data. As with Currie's Concepts, simplified conceptual examples are presented for the situation where decisions are based on individual measurements. The situation with respect to decisions using entire datasets is discussed as well, although the answers are not so clear-cut at this time. The issues involved have to do with whether to use conventionally censored data or uncensored data, and technical statistical details of dealing with the actual data distributions involved, since the conventional models used are appropriate for distributions of actual concentrations, but distributions of measurements are more complicated. Research needs are identified.

The final section is about the use of uncensored data. In additional to the technical issues involved, there are non-technical issues related to public relations and risk communication and to accreditation requirements.

References

1. ISO/IEC 17025:2005, *General Requirements for the Competence of Testing and Calibration Laboratorie*s, International Organization for Standardization, Geneva, 2005.
2. L. A. Currie, *Anal. Chem.*, 1968, **40**, 586–593.
3. LQAP Policy Document, Module 2A, American Industrial Hygiene Association, Cincinnati, OH, www.aiha.org, accessed 16 December 2008.
4. D. M. Rocke and S. Lorenzato, *Technometrics*, 1995, **37**, 176–184.
5. ASTM D6091, *Standard Practice for 99%/95% Interlaboratory Detection Estimate (IDE) for Analytical Methods with Negligible Calibration Er*ror, ASTM International, West Conshohocken, PA, 2003.

6. C. B. Davis and R. J. McNichols, *Ground Water Monit. Rem.*, Vol. 14, 1994, **1**, 148–158.
7. C. B. Davis and N. E. Grams, presentation at the US EPA's 25th Annual Conference on Managing Environmental Quality Systems, Austin, TX, 2006.
8. A. Hubaux and G. Vos, *Anal. Chem.*, 1970, **42**, 849–855.
9. C. A. Clayton, J. W. Hines and P. D. Elkins, *Anal. Chem.*, 1987, **59**, 2506–2514.
10. R. D. Gibbons and D. E. Coleman, *Statistical Methods for Detection and Quantification of Environmental Contamination*, John Wiley & Sons, New York, 2001.
11. J. A. Glaser, D. L. Foerst, G. D. McKee, S. A. Quave and W. L. Budde, *Env. Sci. Technol.*, Vol. 15, 1981, **1**, 1426–1435.
12. US Environmental Protection Agency, *Fed. Reg.*, 2003, **68**, 11770–11793, March 12 and supplement.
13. *Final Report of the Federal Advisory Committee on Detection and Quantitation Approaches and Uses in Clean Water Act Programs*, US Environmental Protection Agency, Washington, DC, 2007, www.epa.gov/waterscience/methods/det/faca/final-report-200712.pdf, accessed 12 February 2009.
14. ASTM D6512, *Standard Practice for Interlaboratory Quantitation Estimate*, ASTM International, West Conshohocken, PA, 2007.
15. ASTM International, *New Proposed Practice for Performing Detection and Quantitation Estimation and Data Assessment Utilizing DQCALC Software, Based on ASTM Practices D6091 and D6512 of Committee D19 on Water*, ASTM International, West Conshohocken, PA, 2008.
16. *Multi-Agency Radiological Laboratory Protocols Manual (MARLAP)*, 2004, NUREG-1576, EPA 402-B-04-001, www.epa.gov/marlap/manual.htm, accessed 16 December 2008.
17. D. M. Crumbling, C. Groenjes, B. Lesnik, K. Lynch, J. Shockley, J. Van Ee, R. Howe, L. Keith and J. McKenna, *Environ. Sci. Technol.*, 2001, **35**, 404A–409A.
18. C. B. Davis and T. E. Gran, presentation at the Beryllium Health and Safety Committee Workshop, Aberdeen, MD, April 2008.
19. D. Coleman and L Vanatta, *Am. Lab.*, June/July 2007, 24–25.
20. ISO 5725-6, *Accuracy (Trueness and Precision) of Measurement Methods and Results*, International Organization for Standardization, Geneva, 1994.
21. *Guide to the Expression of Uncertainty in Measurement (GUM)*, International Organization for Standardization, Geneva, 1995.
22. N. A. Leidel, K. A. Busch and J. R. Lynch, *Occupational Exposure Sampling Strategy Manual*, National Institute for Occupational Safety and Health, Cincinnati, OH, 1977, Publ. 77–173.
23. R. O. Gilbert, *Statistical Methods for Environmental Pollution Monitoring*, Van Nostrand Reinhold, New York, 1987.
24. D. R. Helsel, *Nondetects and Data Analysis: Statistics for Censored Environmental Data*, John Wiley & Sons, Hoboken, NJ, 2005.

25. C. B. Davis, *presentation at the Joint Statistical Meetings,* Seattle, WA, 2006.
26. D. K. Bhaumik and R. D. Gibbons, *Technometrics,* 2006, **48**, 112–119.
27. A. Singh, A. K. Singh and R. J. Iaci, *Estimation of the Exposure Point Concentration Term Using a Gamma Distribution,* US Environmental Protection Agency, Washington, DC, 2002, EPA/600/R-02/084.
28. *USEPA Contract Laboratory Program National Functional Guidelines for Inorganic Data Review,* US Environmental Protection Agency, Washington, DC, 2004, EPA 540-R-04-004, www.epa.gov/superfund/programs/clp/guidance.htm, accessed 17 December 2008.

CHAPTER 9

Applications, Future Trends, and Opportunities*‡

GEOFFREY BRAYBROOKE[a] AND PAUL F. WAMBACH[b]

[a] Supervisory Industrial Hygienist, IH Field Services Program, US Army Center for Health Promotion and Preventive Medicine, Aberdeen Proving Ground, MD 21010, USA; [b] Office of Worker Safety and Health Policy, US Department of Energy, 1000 Independence Avenue, SW, Washington, DC 20505, USA

Abstract

Currently beryllium exposure monitoring uses measurement technology developed for metals in general. The very low levels that beryllium exposure control programs are attempting to achieve are leading to innovative analytical methods and could drive innovative exposure monitoring strategies as well. Current strategies generally aim to use air monitoring results for two purposes: 1) to estimate individuals' actual exposures and 2) understand the causes and sources of exposure. This strategy dictates collecting samples from small air volumes. The samples need to be collected from workers breathing zones with wearable monitors to reflect the workers actual exposure and the duration of sampling is limited to a single work shift or shorter periods so that the results

---

*This article was prepared by US Government employees as part of their official duties. The US Government retains a nonexclusive, paid-up, irrevocable license to publish or reproduce this work, or allow others to do so for US Government purposes.

‡*Disclaimer:* Mention of company names or products does not constitute endorsement by the US Army or the US Department of Energy (DOE). The findings and conclusions in this paper are those of the authors and do not necessarily represent the views of the US Army or DOE.

Beryllium: Environmental Analysis and Monitoring
Edited by Michael J. Brisson and Amy A. Ekechukwu
© Royal Society of Chemistry 2009
Published by the Royal Society of Chemistry, www.rsc.org

can be related to jobs or tasks that are the likely cause of the exposure. Strategies that accomplish these two goals with separate measurements might allow for larger air volumes or more sensitive particle counting instruments. Beryllium risk management also utilizes surface sampling to understand the distribution of contamination and to determine whether skin protection is effective. Recent developments in dermal monitoring are beginning to be used for beryllium exposure control as well. Surface wipe sampling methods developed as a quality check for cleaning are proving to be less well suited for characterization of surfaces with settled dust. Vacuum sampling may be better suited for this problem if agreed-upon methods for interpretation can be reached. Technology that could be used for point-and-shoot instruments has been demonstrated but would require significant investment to develop.

9.1 Introduction

Beryllium measurement technology utilizes methods developed for metals in general.[1] Exposure monitoring and surface sampling for beryllium have benefited from the gradual improvement in the sampling and analytical instruments used for other metals. However, recent health studies point to the need to further reduce beryllium occupational exposure limits (OELs) to levels less than those for many other metals, calling into question whether existing sampling and analytical methods will answer the questions that were the reason for making the measurement in the first place. In this chapter, we look at the objectives for measuring beryllium with the assumptions that these will be important drivers for the development of new technology. Monitoring is expensive and reducing costs will always be a driver for improved methods. We then look at newly available methods and discuss their ability to meet objectives. Objectives that are likely to remain unmet are gaps that will require research to fulfill.

9.2 Monitoring

There are two broad categories of monitoring objectives: 1) to estimate individuals' actual exposure levels; and 2) to determine the sources and causes of exposures. Often a measurement will be made to help meet both objectives.

9.2.1 Baseline Monitoring

"Baseline" monitoring is a strategy for assessing an individual's actual exposure level.[2] It is routine ongoing monitoring of groups of workers who share similar exposure determinants such as job, task, and location. In most settings, only a small percentage of all work shifts can be monitored. Monitoring should be statistically planned and conducted with sufficient frequency to achieve a specified level of confidence that exposures are being maintained below OELs.

Measurements are considered to be representative of exposure for every member of the group and used to estimate metrics such as percent exceeding the OEL and confidence intervals. In some situations, every worker can be monitored during every shift and each individual's exposure will be completely determined.

A key data quality objective of baseline monitoring is to minimize the chance of a false negative result, *i.e.* concluding that individuals are not being overexposed when in fact they are. Minimizing false negative results requires a monitoring method sensitive enough to detect exposures of at least an order of magnitude below OELs. Since the distributions of occupational exposures in a group tend to be highly skewed, levels of more than a factor of 10 or 20 below an OEL are an indication that exposures above the OEL may be occurring with an unacceptably high frequency. When the potential for significant exposure exists, baseline monitoring must include enough work shifts to detect infrequent overexposure events. Hypothesis testing statistics provide methods for assuring exposures are being monitored often enough.

Individuals' exposure levels will be determined in part by their distance from sources of beryllium aerosol generation and in part by the dust generating or disturbing tasks they perform. Wearable personal samplers were developed in the 1960s to integrate these two sources of exposure. They were shown to help minimize false negative conclusions caused by older methods that tended to underestimate exposure from tasks. However, adoption of these methods for beryllium monitoring were delayed until the 1980s in part because of the inability of older analytical instruments to quantify relevant levels in the smaller air samples collected by personal sampling equipment. A key objective for monitoring exposure to airborne beryllium is the ability to collect and analyze aerosols from the worker's breathing zone with wearable samplers.

9.2.2   Compliance Monitoring

"Compliance" monitoring is intended to quickly identify exposures above OELs by focusing on suspected high exposure jobs and tasks.[2] This is often the first step when evaluating an operation that has not been monitored previously. It is also used when there are inadequate resources to establish a baseline monitoring program. Since even a single measurement above an OEL is convincing evidence that OELs are being exceeded too frequently, compliance monitoring can be a very cost-effective method of identifying working conditions that require immediate corrective action.

The term compliance monitoring comes from its use by inspectors with the US Occupational Safety and Health Administration (OSHA) to quickly identify non-compliant workplaces in the limited time available for an inspection. A single result will provide the legal basis for a fine and to require corrective actions. In this setting, a key data quality objective is to minimize the chance of false positive results, *i.e.* the risk of imposing a fine for non-compliance with a regulatory OEL when it hasn't been exceeded. The chance of a false positive result can be minimized with sampling and analytical methods with a high degree of precision near the OEL.

9.2.3 Diagnostic Monitoring

"Diagnostic" monitoring is aimed primarily at determining the source or cause of exposure.[2] This is often deterministic rather than statistically planned and can employ a variety of methodologies. It is usually intended to characterize emissions from tasks, tools, or process equipment rather than an individual's exposure.

Surface wipe sampling to characterize the distribution of contamination in a facility is diagnostic sampling that will be statistically planned. Surface wipe sampling is the most common form of diagnostic sampling for beryllium, since knowledge of the location of contamination is critical for the protection of workers who repair, maintain, and remodel facilities, utilities, and equipment.

9.2.4 Exposure Monitoring

Personal exposure monitoring for relatively short periods of time, such as for a single work shift or for the duration of a task within a shift, can also provide information on the source or cause of an exposure. Sampling over longer periods of time, such as a week or month, loses information on associations between measured levels and determinants of the levels. In addition, OELs are usually applied to a single shift to control dose rates to safe daily intake levels and to assure action is taken to prevent exposure at the earliest indication of hazardous conditions. The ability to use a single sample in both baseline and diagnostic strategies creates a strong preference for eight-hour samples over longer term samples (*e.g.* 40 hours) that might provide a larger sample and allow quantitation at lower levels. The battery-powered personal sampling pumps currently on the market can collect $1-2\,m^3$ of air in an eight-hour shift.

9.2.5 Future Trends

There are several factors that have the potential to affect beryllium monitoring in the future. As noted in Chapters 1 and 2, there is a tendency towards lower exposure limits, often with specified size fractions (US DOE, US OSHA, California OSHA, ACGIH®). There is an ongoing need tor obtain faster results. Further research is expected to improve our understanding of health effects, particularly the influences of particulate size and number on respiratory exposures, the role of dermal exposure, and the role of the different chemical species in which beryllium occurs.

9.3   Air Sampling

As noted in previous chapters, the American Conference of Governmental Industrial Hygienists (ACGIH®) has published a Notice of Intended Change (NIC) to the Threshold Limit Value (TLV®) for beryllium to $50\,\text{ng}\,\text{m}^{-3}$ as an eight-hour time-weighted average (TWA).[3] This is to be measured as the inhalable fraction of particulate. This new limit was derived by applying a safety factor of about 10 to levels that have been associated with chronic beryllium disease (CBD). It is consistent with the 1998 US Environmental Protection Agency (EPA) reference concentration (RfC) for beryllium of $20\,\text{ng}\,\text{m}^{-3}$ as a 24-hour limit.[4] It is unlikely that regulatory or other author-itative OELs for beryllium will be significantly different from this.

As noted above, statistical confidence intervals are used to determine the frequency of baseline monitoring. The greater the distance between a mea-surement result and the OEL, the more confidence one has that the OEL is not being exceeded. With an OEL of $50\,\text{ng}\,\text{m}^{-3}$ and the use of personal sampling equipment capable of collecting a $1–2\,\text{m}^3$ sample, a Reporting Limit (RL) of 1– 5 ng per sample is an achievable data quality objective. Reporting limits are typically associated with quantitation limits a factor of 3–10 less, or about 0.5 ng per sample.

The ACGIH® NIC also includes a 15-minute Short Term Exposure Limit (TLV-STEL) of $200\,\text{ng}\,\text{m}^{-3}$, also measured as the inhalable fraction. This OEL is intended to support interpretation of diagnostic monitoring results of short duration tasks. However, a 15-minute sample collected using current personal sampling equipment would require analytical methods capable of reporting about 6 ng per sample to demonstrate non-compliance. Again, a data quality objective for demonstrating compliance would be a reporting limit of about one tenth this level, or 0.6 ng per sample.

ACGIH® is progressively moving towards defining airborne particulate TLVs in terms of defined respirable, inhalable, or thoracic size fractions.[3] International Standard ISO 7708 also includes definitions of respirable, inhalable, and thoracic size fractions.[5] While ACGIH® is currently considering defining its TLV® in terms of the inhalable fraction, future research into the effects of particle inhalation may focus research and regulatory attention on other size fractions. However, the sampling devices and sampling rate in actual use are inconsistent between countries, and much work is required to ade-quately define the fractions to incorporate real-world variables such as wind velocity. There is also a possible need to define and measure an ultrafine size fraction.[5] Devices and methods for inhalable and respirable fraction sampling are discussed in detail in Chapter 2. Future consistent use of a single device with a single sampling rate for each fraction of interest would be desirable to allow comparisons between different studies and sites.

These three objectives – sampling air from an individual's breathing zone, sampling for short time periods, and sampling size selected fractions of total aerosol – all make air sampling and analytical methods that can achieve lower reporting limits desirable.

Improvements in pump design and in the power-to-weight ratio of pump batteries may make higher sampling rates possible, thus collecting larger masses of particulate in a given period. Flow-rates may, however, be constrained by needs for size selective sampling.

9.4 Analytical Methods

The analytical method of fluorescence[6] and inductively coupled plasma mass spectrometry (ICP-MS),[7] discussed in Chapters 6 and 7, are already able to achieve reporting limits of a few nanograms per sample. These have yet to be widely adopted by environmental chemistry laboratories, but there does not appear to be any fundamental barrier to their adoption beyond the cost of the equipment.

There would be further advantage to extending the analytical capability down to a few picograms per sample. Trace levels in ambient air would result in detection of beryllium at this level in most air samples and probably most surface samples as well. Almost every measurement will produce a result, and analytical uncertainty would cease to be an important contributor to the overall uncertainty in estimates of an individual's exposure.

9.5 Speciation

Exposure to different beryllium species have been associated with a range of acute and chronic lung and skin health effects.[8] Solubility is the key determinant of whether the effect will be acute or chronic. Recent epidemiology studies have reported on the association between CBD and exposure to poorly soluble aerosols generated by beryllium metal, alloy and ceramic utilization, and fabrication operations. CBD is prevalent at exposure levels much lower than those associated with acute disease and its prevention is the primary concern of health protection programs today. Exposures to the more soluble forms of beryllium that caused acute disease occurred in plants that processed ore. Only one primary production plant remains in business in the US today, further limiting interest in characterizing the more soluble aerosols produced by these operations. However, it is now apparent that dusty construction activities, such as abrasive blasting, can result in exposure to naturally occurring beryllium silicates at levels well above OELs.[9] Studies of workers involved in mining and ore processing, and in gem stone polishing, have not found cases of CBD that could be attributed to exposures to beryllium silicates; however, the studies were too small to rule out the possibility that these findings are due to chance.[10,11] The resulting uncertainty creates an interest in exposure monitoring capable of distinguishing between the silicate and the metallic and oxide forms of beryllium. Even more recently, a report has been published on a case of CBD associated with a beryllium aerosol that is the byproduct of primary aluminium production.[12] The aerosol is thought to contain a mixture of predominantly soluble beryllium fluoride salt and smaller amounts of poorly soluble beryllium oxide. Again the ability to differentiate between these two species would be desirable to gain a better understanding of the risks in this industry. A conventional approach to this problem is serial dissolution with increasingly acidic solvents. One attempt has been made to develop a sequence of analyses that does this,[13] but it has not been successfully replicated to date.

9.6 Making Use of Censored Data

A problem that is commonly encountered in beryllium surveys is that a large proportion of each dataset consists of results reported as being below the laboratory reporting limit (commonly known as "censored data" or "non-detects"). If it can be assumed that the distribution is log-normal, as is often true for airborne exposure data, maximum likelihood estimates can be used to estimate the confidence interval for an exceedance fraction and other parameters by assuming non-detected values would fall on a line fitted to the detected values. This method has been applied for the statistical analysis of datasets with up to 80% censored data. When the use of a log-normal model cannot be justified, product limit estimates and other non-parametric order statistics can be used to estimate the confidence interval for an exceedance fraction; however, this requires larger numbers of samples to achieve the desired level of confidence.[14] The US Department of Energy has developed free software to perform these calculations.[15]

In at least one environmental chemistry application, the analysis of radio-isotopes in soil, reporting limits have been replaced with uncertainty statements. This approach seems compatible with ISO 17025.[16] When a laboratory establishes reporting limits, it must make assumptions about its customers' data quality objectives and the default assumption is that customers desire a high degree of precision to minimize the chance of a false positive result. If the customer's objective is to minimize chances for false negative results, he or she might be willing to tolerate much lower levels of precision to gain information on the distance between a measurement and an OEL. Reporting all results, even a negative number, with an uncertainty statement (*i.e.* a confidence interval) allows the customer to use all the available information to contribute to a decision.

9.7 Dermal Sampling

Some recent studies have included assessments of skin exposure to determine if it is associated with risk for beryllium sensitization.[17] If an association is established, it is possible that routine skin monitoring would be instituted. Currently skin monitoring is an adaptation of surface monitoring methods. Wetted surface sampling media are used to wipe hands, face, and neck and results are used as an indication of the degree of protection being provided by clothing and other personal protective equipment. Adhesive tape is also used for skin sampling.

As skin sampling is becoming more common, wearable media such as patches and gloves are being developed to provide a better indication of skin exposure. Thin cotton gloves and other media being considered are likely to create digestion challenges in the future. Development of standard media and methods for dermal sampling would be desirable to allow comparisons among different studies and sites. To this end, technical committees within ASTM International and the International Standards Organization (ISO) are currently developing dermal sampling standard guidance and/or methods.

9.8 Surface Dust Sampling

Surface sampling for beryllium contamination utilizes methods originally developed for lead sampling, which is a major public and occupational health hazard. With current wipe media, dry wiping appears to have a much lower collection efficiency than the adequate levels that are achievable with wet wiping, but is necessary for surfaces that are damaged by moisture. A "tacky' dry wipe has been invented that may become a commercial product. It is intended for cleaning surfaces rather than collecting samples, and its compatibility with sample preparation and analytical methods has not yet been evaluated.[18]

Evaluations of lead contamination are using vacuum sampling in addition to surface wipe sampling as a method capable of providing information on both surface loading levels and the concentration of lead in materials collected. Wipe sampling with pre-weighed filters has also been used to determine concentration. The spatial distribution of concentrations provides information on source, since concentration is highest near the source. Because trace levels of beryllium are present in the environment, knowing whether the source of contamination is soil, building materials or a process can be important in many settings. Analyses of both the total mass of collected material and the beryllium fraction can help answer that question. Another approach is to determine the ratio between beryllium and another metal such as aluminium in the sampled dust to compare it with the ratio in ambient soil concentrations.

9.9 Real-Time/Near Real-Time Measurement

9.9.1 Research History

For beryllium, it would be very desirable to have a real-time or near-real-time technology that ideally could both monitor air and analyze surface wipes, with a reporting limit similar to fixed laboratory methods, which would be easy to operate and maintain, inexpensive, and produce minimal or no liquid hazardous waste. Real-time instruments would save time and money by avoiding laboratory turnaround delays, reducing overall sampling and analytical costs, and providing instant feedback for worker training, awareness, and protection. With beryllium being a small niche market for monitoring instrumentation, it would be difficult for a commercial manufacturer to justify the investment in developing an instrument for this application alone. Therefore, real-time technology would ideally be suitable for all species of beryllium and for other metals.

For example, X-ray fluorescence (XRF) instruments are widely used as field-portable survey meters for lead and many other metals, but do not work for beryllium. Used as a survey meter, the XRF provides a "pass/fail" indication of whether a metal is present at a given level. These instruments can also be calibrated to provide useful field quantitation of metal content on filters, wipes, and soil samples. Due to the transparency of beryllium to X-rays, XRF is not viable for measuring beryllium. However, it would be desirable to have an instrument that could serve a similar role for beryllium.

Prior to 2002, several US Department of Energy sites and instrument manufacturers had independently experimented with various technologies. These were presented at the First Symposium on Beryllium Particulates and Their Detection organized by the Beryllium Health and Safety Committee (BHSC) in 2002.

One technology considered was laser induced breakdown spectroscopy (LIBS). This method produces no liquid waste, and one supplier offered both air and wipe models. However, one unit required a cart rather than being truly portable. A microwave induced plasma spectrometer (MIPS) was developed in 2002 and was commercialized by 2005. An anodic stripping voltammetry (ASV) instrument was developed by one manufacturer that was portable but generated liquid waste. An aerosol time-of-flight mass spectrometer (ATOFMS) required a cart and was expensive. Also presented were an aerosol-focusing LIBS and a surface-enhanced Raman spectrometer (SERS). As later noted by Brisson et al.,[19] all of these technologies lacked comprehensive validation and the publication of standard methods for their use as a prerequisite for laboratory certification.

As a result, the DOE Network of Senior Scientists and Engineers established a Beryllium Advanced Technology Assessment Team with representation from various DOE sites and other federal agencies to identify needs and make recommendations. Draft performance and evaluation criteria were developed for both air monitoring and surface wipe analysis equipment.

By the time of the Second Symposium on Beryllium Particulates and Their Detection in 2005, a commercial product had been developed to perform fluorescence analysis as described in Chapter 7. Development work had also continued on the aerosol-focusing LIBS and the Raman spectrometer mentioned above.

As of 2005, the Beryllium Advanced Technology Assessment Team saw a need to refine the criteria, especially in terms of what needed to be measured – mass, particle number, surface area, or some combination thereof, in total beryllium or select species. Matrix compatibility and interference issues needed to be addressed. Coordinated technology development would be desirable. However, as yet, there has been a lack of funding to support Team activities and technology development, and the important toxicological parameters that will define what exactly needs to be measured are still unclear.[20]

9.9.2 Future Research Directions

Day-to-day health risk management involves predicting the degree of risk associated with planned work and recommending controls to prevent excessive exposure. Exposure monitoring results are used to validate that the controls were effective or quickly correct any deficiencies. There is ongoing research aimed at improving the initial prediction through the use of exposure modeling. The research utilizes real-time exposure monitoring instruments and synchronized video recording to quantify emissions associated with material characteristics, the energy associated with process equipment, and air movement in the workroom. While the goals of this research are development of mathematical models that are able to reliably predict exposure from knowledge of materials, tools and ventilation, it is also providing new knowledge on the relationship between aerosol measurements and an individual worker's exposure. This new knowledge may also enable development of reliable calibration factors that would allow use of an easy-to-measure parameter, such as particle number, as a reliable surrogate for directly measuring beryllium exposure.

New models of light scattering particle counters are greatly improved and relatively inexpensive. They are wearable, can be used for size selective sampling, and are sensitive enough to provide particle counts in ambient background conditions. However, interpretation of particle count measurements requires knowledge of the percentage of the total aerosol that is the contaminant of concern and the particle size distribution to convert particle count into a value that can be compared to a mass-based OEL. Currently gaining this knowledge requires specialized expertise and is specific to a particular operation. If calibration factors can be generalized across many similar workplaces, then the effort is worthwhile but more difficult to justify if the calibration is unique to each workplace or worst-case each operation. Calibration factors have been developed for coal mining.[21] It is possible that calibration factors will be developed for metalwork that can be applied to beryllium work, or that exposure modeling research will provide new knowledge will make it possible to adjust calibration factors based on readily available information on the materials and tools being used in an operation.

Unlike beryllium, where short-term personal monitoring provides both baseline and diagnostic information, radiation dosimeters are long-term monitors read monthly, quarterly, or even annually. An array of other instruments, area alarms, portal alarms, survey meters, pocket meters, and ring meters are used for day-to-day radiation protection and to identify sources of exposure. Inexpensive wearable particle counters have the potential to be integrated into new beryllium exposure monitoring strategies in which they provide semi-quantitative meter and alarm functions for day-to-day risk management. In this role, establishing calibration factors could be simplified by using conservative estimates of the aerosol's percentage beryllium and particle size. Definitive exposure information could be provided by current personal exposure monitoring equipment with sampling periods long enough to provide quantifiable results of size selective samples.

References

1. *Method 7300*, Elements by ICP, in *NIOSH Manual of Analytical Methods*, ed. P. C. Schlecht and P. F. O'Connor, US National Institute for Occupational Safety and Health, Cincinnati, OH, 4th edn, 2003, issue 3, www.cdc.gov/niosh/nmam/, accessed 12 February 2009.

2. J. Mulhausen, J. Damiano and E. L. Pullen, Further information gathering, in *A Strategy for Assessing and Managing Occupational Exposures*, ed. J. Ignacio and W. Bullock, American Industrial Hygiene Association, Fairfax, VA, 2006.

3. *TLVs$^{®}$ and BEIs$^{®}$*, American Council of Governmental Industrial Hygienists, Cincinnati, OH, 2008.

4. *Toxicological Review of Beryllium and Compounds*, US Environmental Protection Agency, Washington, DC, 1998, EPA/635/R-98/008, www.epa.gov/ncea/iris/toxreviews/0012-tr.pdf, accessed 4 September 2008.

5. L. C. Kenny, *Appl. Occ. Env. Hyg.*, 2000, **15**, 68–71.

6. Method 7704, Beryllium in Air by Field Portable Fluorometry, and Method 9110, Beryllium in Surface Wipes by Field Portable Fluorometry, in *NIOSH Manual of Analytical Methods*, ed P. C. Schlecht and P. F. O'Connor, National Institute for Occupational Safety and Health, Cincinnati, OH, 4th edn, 1994–2006, www.cdc.gov/niosh/nmam/, accessed 12 February 2009.

7. ASTM D7439, *Standard Test Method for Determination of Elements in Airborne Particulate Matter by Inductively Coupled Plasma—Mass Spectrometry*, ASTM International, West Conshohocken, PA, 2008.

8. J. H. Sterner and M. Eisenbud, *Arch. Ind. Hyg. Occup. Med.*, 1951, **4**, 123–151.

9. P. K. Henneberger, S. K. Goe, W. E. Miller, B. Doney and D. W. Groce, *J. Occup. Environ. Hyg.*, 2004, **1**, 648–659.

10. D. Deubner, M. Kelsh, M. Shum, L. Maier, M. Kent and E. Lau, *Appl. Occup. Environ. Hyg.*, 2001, **16**, 579–592.

11. R. Wegner, R. Heinrich-Ramm, D. Nowak, K. Olma, B. Poschadel and D. Szadkowski, *Occup. Environ. Med.*, 2000, **57**, 133–139.

12. O. A. Taiwo, M. D. Slade, L. F. Cantley, M. G. Fiellin, J. C. Wesdock, F. J. Bayer and M. R. Cullen, *J. Occup. Environ. Med.*, 2008, **50**, 157–162.

13. A. Profumo, G. Spini, L. Cucca and M. Pesavento, *Talanta*, 2002, **57**, 929–934.

14. E. L. Frome and P. F. Wambach, *Statistical Methods and Software for the Analysis of Occupational Exposure Data with Non-Detectable Values*, prepared by Oak Ridge National Laboratory for the US Department of Energy, 2005, ORNL/TM-2005-52, www.hss.energy.gov/HealthSafety/IIPP/sand/ORNLTM2005-52.pdf, accessed 8 September 2008.

15. Statistical Analysis of Non-detect Data (SAND), US Department of Energy, Office of Health and Safety www.hss.energy.gov/HealthSafety/IIPP/sand/, accessed 8 September 2008.

16. ISO/IEC 17025:2005, *General requirements for the competence of testing and calibration laboratories*, International Organization for Standardization, Geneva, 2005.
17. G. A. Day, A. Dufresne, A. B. Stefaniak, C. R. Schuler, M. L. Stanton, W. E. Miller, M. S. Kent, D. C. Deubner, K. Kreiss and M. D. Hoover, *Ann. Occup. Hyg.*, 2007, **51**, 67–80.
18. *Flypaper for Particulate: New Y-12 Tack Cloth Leaves No Sticky Residue*, News Release, US DOE Y-12 National Security Complex, Oak Ridge, TN, 19 February 2007, www.y12.doe.gov/news/release.php?id=61, accessed 12 December 2008.
19. M. J. Brisson, K. Ashley, A. B. Stefaniak, A. A. Ekechukwu and K. L. Creek, *J. Environ. Monit.*, 2006, **8**, 605–611.
20. M. J. Brisson and K. Creek, *Criteria for Real-Time Beryllium Monitoring Equipment*, presented at the 2nd Symposium on Beryllium Particulates and Their Detection, Salt Lake, Utah, November 2005, www.sandia.gov/BHSC
21. S. J. Page, J. C. Volkwein, R. P. Vinson, G. J. Joy, S. E. Mischler, D. P. Tuchman and L. J. McWilliams, *J. Environ. Monit.*, 2008, **10**, 96–101.

Subject Index

www.ingramcontent.com/pod-product-compliance
Lightning Source LLC
Chambersburg PA
CBHW031950180326
41458CB00006B/1687